CATCHING CRYPTIDS

CATCHING CRYPTIDS

THE SCIENTIFIC SEARCH FOR MYSTERIOUS CREATURES

KIM LONG

Illustrated by NICOLE MILES

RP | KIDS
PHILADELPHIA

Running Press Kids
Hachette Book Group
1290 Avenue of the Americas, New York, NY 10104
www.runningpress.com/rpkids
@runningpresskids

First Edition: May 2025

Published by Running Press Kids, an imprint of Hachette Book Group, Inc. The Running Press Kids name and logo are trademarks of Hachette Book Group, Inc.

The Hachette Speakers Bureau provides a wide range of authors for speaking events. To find out more, go to www.hachettespeakersbureau.com or email HachetteSpeakers@hbgusa.com.

Running Press books may be purchased in bulk for business, educational, or promotional use. For more information, please contact your local bookseller or the Hachette Book Group Special Markets Department at Special.Markets@hbgusa.com.

The publisher is not responsible for websites (or their content) that are not owned by the publisher.

Print book cover and interior design by Mary Boyer

Library of Congress Cataloging-in-Publication Data
Names: Long, Kim, author.
Title: Catching cryptids: the scientific search for mysterious creatures/Kim Long, Nicole Miles.
Description: First edition. | Philadelphia: Running Press Kids, 2025. | Audience: Ages 8-12. | Summary: "A STEM-based, fun fact-filled, quirky cryptozoology book for fans of *The Book of Mythical Beasts and Magical Creatures*"—Provided by publisher.
Identifiers: LCCN 2024028481 (print) | LCCN 2024028482 (ebook) | ISBN 9780762485758 (hardcover) | ISBN 9780762485765 (epub)
Subjects: LCSH: Cryptozoology—Juvenile literature. | Animals, Mythical—Juvenile literature.
Classification: LCC QL88.3 .L64 2025 (print) | LCC QL88.3 (ebook) | DDC 001.944—dc23/eng/20240624
LC record available at https://lccn.loc.gov/2024028481
LC ebook record available at https://lccn.loc.gov/2024028482

ISBNs: 978-0-7624-8575-8 (hardcover), 978-0-7624-8576-5 (ebook)

Printed in Guangdong, China

1010

10 9 8 7 6 5 4 3 2 1

CONTENTS

WELCOME LETTER

Hello, there! I imagine you picked up this book because, like me, you're a fan of cryptids. Or perhaps you just want to know more about these mysterious creatures. Either way, let's start with the basics. What is a cryptid?

Cryptids are animals whose existence is disputed or not yet substantiated by science.

You've probably heard of a few, like the Loch Ness monster and Kraken.

Well, did you know there are people who make it their life's goal to find these mysterious creatures? Cryptozoology is the study of "mystery" or "hidden" animals that witnesses say they've seen but have not been proven to exist, and a cryptozoologist is someone involved in the studying of and/or searching for these animals.

This is where we come in. Using the science and technology discussed in this book, we'll join the search and hopefully, finally, locate some of these cryptids! A word of caution, though: proving a cryptid exists will not be easy. For starters, merely *seeing* one of these animals isn't enough to prove it exists—not when we can't trust our eyes to be 100 percent accurate.

You see, scientists studying the accuracy of eyewitnesses have explained that our brains often play tricks on us. When we look at an object, our brain takes what is being reported by our senses and fills in the blanks to complete the image. The process is automatic— we don't even know it's happening. This means that, while we might 100 percent believe we saw a black-and-white cow run up a tree, the truth might be different. Very different. Lighting, distance, and how long we looked at the object play a role in accuracy, as does our emotional state. Were we anxious? Afraid? Distracted? Surprised? If so, our ability to accurately report what we saw decreases.

In other words, to prove a cryptid isn't the result of our trick-playing brain or a figment of our imagination, we have to do more than spot it lurking about. We need to collect scientific evidence of its existence. Only at that point will we be able to determine once and for all what people are seeing.

In the past, cryptid-hunting expeditions depended a lot on luck. Cryptozoologists trekked to an often hard-to-reach or remote location where someone had reported seeing a creature, hoping they'd be lucky enough to see it, too.

Well, we're not going to rely solely on luck for our search. Technological advances have made a huge impact on wildlife science. With cutting-edge tech, scientists can more easily explore previously hard-to-reach territory—like rugged grasslands, the depths of the ocean, and dense jungle. Innovative technology has also helped scientists gather intel on creatures notoriously difficult to find—like nocturnal animals that aren't out and about during the day.

Applying this tech to our missions will be a game changer! We'll be able to learn more about the cryptid and its habitat *before* we make our trek into the field. In essence, by using science to pinpoint a cryptid's potential home, we'll improve our chances of finding it.

What tech we use will depend on the cryptid, of course. Underwater robots aren't going to be much help in the desert, and unless scent-detection dogs are secretly fantastic swimmers, they're not going to be useful in the ocean. So, in order to select the appropriate (or most useful) tech, we'll first have to get to know our cryptids a little better. Then we can determine which technology lends itself to our mission and develop a search protocol.

Sound good?

Then what are you waiting for? Turn the page and let's get to it!

INTRODUCTION:

ARE THERE REALLY CRYPTIDS TO CATCH?

I sense your skepticism. While you're excited about the idea of finding a cryptid, you're also hesitant. You can't help but wonder, "If cryptids are real, wouldn't they have been caught by now?"

Good question.

Short answer: not necessarily.

Animals, it turns out, are really good at hiding.

Anyone who has ever visited a wildlife park, farm, or zoological garden has probably had the experience of not being able to find an animal in its "designated" area. We'll peer into the animal's space from every possible angle, hoping to glimpse a tail, a paw, or something that will guide us to the animal's location.

Our frustrating one-sided game of hide-and-seek is just a peek into an animal's tendency to remain hidden. This elusiveness isn't an accident, either. Study after study shows the same thing: animals deliberately *choose* to conceal themselves from people.

Ever try to pet a squirrel and have it race off in the other direction? Or get close enough to feed a bird on a fencepost only to have it dart into the sky? Even tiny garter snakes slither away if you try to catch them.

Research shows that most animals, if they had their way, would have absolutely nothing to do with people. In an experiment to test animals' reactions to humans, scientists broadcast human voices into areas containing mountain lions, bobcats, opossums, and skunks. Mountain lions abandoned their kill, skunks and opossums foraged less, and bobcats became solely nocturnal.

However, when scientists broadcast frog noises, the animals did not change their behavior.

Some animals want to avoid humans so badly they'll alter their daily routine. Tigers at a national park in Nepal started sleeping during the day to avoid humans walking on trails. And in one study, when badgers were exposed to human voices, they chose to stay in their burrows longer than when exposed to animal sounds. Even pumas kept their distance and slowed their movements when confronted with human voices.

An animal's ability to avoid humans is so great that scientists regularly come across animals thought to have gone extinct because no one had seen them in years.

Like the night parrot. Between 1912 and 1979, the little green Australian bird went unseen, which led scientists to believe it was extinct. In 1979, a possible sighting was reported, but then the parrot disappeared again until the 1990s. Sporadic possible appearances followed, and then in 2015, sound recordings of the bird were captured, confirming its presence.

So, what happened? How could a bird not be seen for nearly 70 years? Turns out the night parrot is quite elusive. It only comes out after dusk and goes to bed before dawn. It also flies low, often sticking to the ground.

Then there's the Fernandina giant tortoise, last seen in 1906 until one was spotted over 100 years later in June 2019. In 2021, scientists confirmed the turtle was the species thought to have gone extinct due to volcanic activity on its home island in the Galápagos.

The night parrot, Fernandina giant tortoise, and other "rediscovered" animals show it is possible for creatures to "hide" for decades.

BUT I THOUGHT YOU WERE DEAD?!

Scientists have confirmed the existence of quite a few species that were once thought to be extinct. Here are some other "rediscovered" animals.

ANIMAL	LAST SEEN	REDISCOVERED	TIME LAPSE
Jerdon's babbler (bird)	1941	2014	73 years
Pinocchio lizard	1960s	2005	45 years
La Palma giant lizard	1500s	2007	500+ years
Kashmir musk deer (small deer with fangs!)	1948	2014	66 years
Black kokanee or kunimasu (salmon)	1935	2010	75 years
Pygmy tarsier	1921	2000	79 years
Ivory-billed woodpecker	1944	2022	78 years
Cuban solenodon or almiquí (mole-like animal)	1890	1974	84 years
Bermuda petrel (bird)	1603	1951	348 years

"But that's different!" you might be thinking. "No one was looking for that tiny bird or old turtle, so of course they missed it. People have been looking for the Loch Ness monster and some of these other cryptids for years. Doesn't that make it more likely they would have been found if they exist?"

Short answer: again, not necessarily.

While it's probably true *you* weren't seeking out the night parrot or the Fernandina giant tortoise, studying habitats, finding new species, and working to protect ecosystems is what some scientists do. It's *all* they do. Yet animals repeatedly evade detection.

Take, for example, the giant panda. In 1869, a French naturalist who had been exploring China brought back with him remains of a black-and-white panda. This was the first time Western countries became aware of its existence, and they were enthralled with the creature. Museums immediately dispatched expeditions to the area in hopes of obtaining a live panda for their exhibits.

They failed. Keep in mind, we're not talking about a small animal here. Giant pandas stand 2 to 3 feet tall at the shoulder when on all fours and 4 to 6 feet tall when on their hind legs. They can weigh up to 330 pounds. Yet, despite their size and the fact that the panda's general location was known, the panda remained undetected by experienced explorers for 60 years, until 1929, when one was found in a tree.

The giant panda isn't the only animal that was able to escape detection by people directly searching for it. Did you know several animals we recognize today *were once considered cryptids*?

CRYPTIDS HAVE BEEN FOUND BEFORE

You've heard of the platypus, right? At one point, people were adamant it couldn't exist. Unbelievable? Well, what would you think if you were given the following description?

- ◆ It's about 2 feet long.

- ◆ It has a bill like a duck.

- ◆ It has a flattened tail like a beaver.

- ◆ It has feet like an otter.

- ◆ It has ½-inch spurs (picture a dog's canine teeth) on its ankles.

- ◆ Oh, and venom shoots out of those spurs.

What? A large duck-beaver thing that shoots venom? Yeah, there's no way *that* could be real.

Which is exactly what Europeans thought long ago. Since the platypus lives only in eastern Australia, no one from Europe had ever seen the creature. In the late 1700s, when European travelers sent the hide of the supposed platypus back home as evidence, they were accused of creating a hoax. Even the European zoologist who eventually examined a complete specimen wondered if someone was playing a trick on him. He used scissors to try and find stitches, thinking someone might have sewn a duck's beak on the body of a beaver-like animal. Ultimately, he concluded the animal was genuine and gave it the name platypus.

A more recent example of an animal losing its "cryptid" classification and becoming recognized as a real animal is the Sri Lanka devil bird. This bird has been part of the country's folklore for centuries. According to legend, it lives in the forest and produces a bloodcurdling shriek at night. Locals believe the bird's cry is an omen predicting death. Few reported ever seeing it, but those who did described a massive bird with a large tail and horns protruding from its head. It had black, glowing eyes that supposedly stared into their souls.

Yikes. A bloodcurdling shriek? Large horns? Black, glowing eyes? Surely if an animal like *that* existed, it would be easy to find.

Nope. It took until 2001, when the unusual spot-bellied eagle-owl was discovered deep inside Sri Lankan forests. Upon further study, the consensus became that this bird was the cryptid rumored to have existed for centuries, because its characteristics matched those from the legend.

☑ **BLOODCURDLING SCREAM?** The spot-bellied eagle-owl utters a mournful mewing scream that rises and then falls in pitch. It also booms low-registered hoots that travel far distances.

☑ **SIZE?** At 20 to 26 inches tall, it is the tallest of the Sri Lankan owls and is the sixth-longest owl in the world.

☑ **NIGHT ACTIVITY?** Like other owls, it is nocturnal and spends its days hidden in the foliage of large trees.

☑ **BLACK EYES?** Its eyes are black.

☑ **HORNS?** The spot-bellied eagle-owl has sweeping sideward-facing "horns" that jut out the top of its head. (These horns are called *plumicorns*, meaning "feathered horns," because they consist of feathers yet look like horns.)

Scientists identify *hundreds* of new species every year, and there are animal discoveries all the time. In 2021, scientists located a new species of chameleon in the jungles of Madagascar. In 2022, zoologists discovered not one, not two, but several new species of sunbirds in Indonesia. In 2023, scientists found a new species of giant spider in the forests of Australia. Similarly, in 2024, scientists published their research regarding a new species of alligator lizard found in Mexico (a multiyear study involving four expeditions!). With advances in technology that now allow easier access to notoriously hard-to-reach or unexplored places, who knows what will be discovered next?

Perhaps the cryptids discussed in this book are out there waiting for us. Or perhaps, while searching for our cryptids, we'll stumble upon a new species. We'll never know unless we get started.

Adventure awaits, so let the search begin!

PART ONE:

AHOY MATEY!: WATER CRYPTIDS

KRAKEN

There may be no sea monster more famous than the Kraken. For good reason, too. Tales of this beast's ocean adventures have been around for more than 800 years.

THE LEGEND

The word *Kraken* comes from Norway, where this creature made its first documented appearance in 1180. Norwegian sailors described a sea monster that wrapped its long arms around ships before engulfing them into the water. Some sailors' reports had the beast creating a maelstrom (mega-powerful whirlpool) that dragged down ships.

Stories of Kraken encounters continued during the hundreds of years that followed. Some witnesses described the Kraken's body as being so large it could be mistaken for an island. Others claimed the

Kraken blackened the sea with a gush of dark liquid and expelled water by its nostrils during an attack.

It wasn't until 1853, when a Danish naturalist encountered a giant cephalopod (marine animal of the mollusk family, like octopus, squid, or cuttlefish) stranded on a beach, that scientific study surrounding the Kraken took off. The naturalist took the animal's "beak" with him, and after examination and 4 years of study, he gave the creature its scientific name: *Architeuthis* ("the chief squid"), also known as the giant squid.

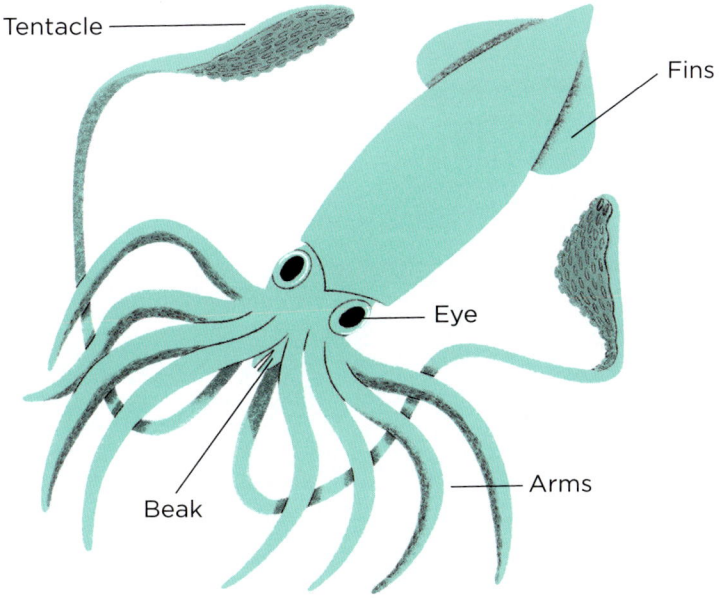

THE GIANT SQUID–KRAKEN CONNECTION

Twenty years later, in 1873, and more than 3,500 miles from Norway, near Great Bell Island in Canada, a fisherman and his son spotted a giant squid. They cut off part of a tentacle, and scientists subsequently estimated the *entire* tentacle must have been 35 feet long (the average length of a school bus) and the creature to which it was attached

60 feet long (the length of a bowling lane) and 5 to 10 feet across. In other words, whatever the tentacle belonged to had to be huge.

This discovery was soon joined by other evidence of giant squid. In 1878, a living giant squid estimated at 55 feet long was found stranded in shallow water in Newfoundland. Then, in 1880, a specimen found on a beach in New Zealand measured 65 feet. In subsequent years, more giant squid specimens were found on shores in varying degrees of decay with visible injuries from, scientists posited, battles with large sperm whales.

GIANT FIGHT!

Sperm whales are one of the largest creatures in the sea. Males can be as long as 80 feet, and they can consume up to 1,000 pounds (that's the size of an adult moose or bear) of food a day. One of their favorite foods? Giant squid. But if we know anything about the giant squid, it's that it doesn't go down without a fight. Scientists have found sperm whales with scars covering their bodies from the giant squid's suckers.

With scientific proof of the giant squid confirmed, one question remained—could an even larger version (e.g., the Kraken) exist?

KRAKEN AS A GIANT SQUID

The giant squid comparison for our Kraken checks a lot of boxes:

- ☑ **LARGE OCEAN ANIMAL?** The "typical" giant squid is estimated at 39 to 53 feet, though the largest known reached more than 60 feet in length. The giant squid's eyes are 10 inches wide (the size of most dinner plates) and are the largest in the animal kingdom.

- ☑ **TENTACLES/ARMS THAT CAN WRAP AROUND SHIPS?** Giant squid have eight arms and two longer tentacles. The suction cups on their arms and tentacles grab hold of objects/marine life (like sperm whales).

- ☑ **FAST IN THE WATER?** Giant squid propel through the water at speeds of 25 miles per hour. They are even capable of rocketing backward.

- ☑ **ABILITY TO BLACKEN THE SEA SURROUNDING THEM?** Like other squid and some octopuses, giant squid can squirt black ink out of a funnel on their body.

One characteristic of the Kraken that doesn't completely fit with the giant squid (or any known creature) is the size. The Kraken was described as being as big as an island and able to wrap its arms around ships. There are reports of squid "attacking" ships as recently as 2003, when one wouldn't detach from a yacht in a sailing race, so that behavior isn't necessarily an issue. But is a giant squid big enough to capsize a ship?

Deep-sea gigantism may hold the answer. This is the scientific theory that creatures living in the ocean depths tend to be larger than their relatives who dwell in shallower water. Several factors, including colder temperatures, food scarcity, the existence of fewer predators, and the fact that creatures living this deep never stop growing, lead to XXXXL sizes.

MY, HOW YOU'VE GROWN!

Some creatures where scientists have noted deep-sea gigantism.

SEA CREATURE	COMMON SIZE	GIGANTISM SIZE
Jellyfish	½ inch to 16 inches	Over 3 feet in diameter (Big Red)
Oarfish	10 feet	30 feet
Isopod	Typically is 2 inches; can be up to 6 inches	About 30 inches
Spider crab	4-7 inches	12½-foot leg span
Giant squid	2-3 feet	40-50 feet

Could the Kraken be a mega-sized giant squid from the depths of the ocean?

If it wasn't so difficult to locate and study the giant squid, we might already have an answer to this question. Unfortunately, scientific examination has been primarily reduced to studying giant squid that have washed up on beaches. These washed-up specimens are hardly in pristine condition. Sometimes, they don't even resemble giant squid.

BLOB OR GLOB(STER)?

It is not unusual for carcasses of sea animals (or parts of dead sea animals) to wash up on shore. But often, by the time a carcass hits the shore, it no longer resembles the animal it once was and instead can only be described as a "large blob" of . . . *something*?

These unidentifiable blobs received their very own definition in 1962: globster.

There have been several notable globsters over the years. Some globsters become so famous they get their own name, like Trunko, the mass that washed ashore on a beach in Margate, South Africa, in 1922. One of the largest globsters washed up off the shore of Tasmania in 1960. It measured 20 feet long and 18 feet wide and had an estimated weight between 10,000 and 20,000 pounds! (Electron microscopes later identified this globster as a whale.)

Why are scientists studying blobs on beaches? Simple. There are no giant squid living in zoos, aquariums, museums, or research labs. Giant squid live so deep in the ocean, they've been nearly impossible to observe much less catch and secure in an artificial environment for further studying.

At its deepest point, the ocean is more than 35,000 feet deep. That's more than 6½ miles. Mount Everest, the tallest mountain on Earth, is around 29,000 feet. Airplanes usually fly between 32,000 and 40,000 feet. If you're ever on a plane (or climbing Mount Everest, if that's your thing), think about how high up you are and then imagine going the same distance down into the ocean. That's *deep*.

Scientists estimate most marine life lives in the *sunlight zone* of the ocean, which is from the surface down to about 650 feet. Between 650 and 3,300 feet, sunlight is faint, the water gets cold (about 39°F), and there's so much pressure that humans cannot survive without mechanical support. This is where the giant squid lives, in the *twilight zone*, which is why it's been so difficult to gather intel on this Kraken suspect.

But there's hope. Advancements in underwater technology have allowed access to the depths of the ocean like never before. Finally, 800 years after the first reports of the Kraken, scientists have the ability to view the giant squid in its natural habitat.

GOING DEEP WITH THE MEDUSA

Submersibles are underwater vehicles that allow scientists a close-up view of deep-sea environments. Human-occupied submersibles usually hold a pilot and one or two scientists and are equipped with lights, cameras, sensors, and robotic arms that can collect samples. Some can dive to depths of more than 21,000 feet, thereby giving scientists access to 99 percent of the ocean floor.

But submersibles are not without their limitations. The deeper you go into the ocean, the darker it gets, and the vehicles use bright white lights to penetrate the darkness. They also utilize propulsion that creates a mix of vibrations and sound. That sound can be extremely noisy—especially in the ocean, where sound travels faster and farther. Bright white lights and noise are likely to scare away deep-sea creatures like the giant squid.

To solve the noise and light problem with underwater vehicles, Dr. Edie Widder and scientists at the Ocean Research & Conservation Association (ORCA) invented the Medusa. It's a stealth camera system that produces minimal noise, and, since no one's driving it, bright lights aren't needed to guide the operator's path. Instead, the Medusa uses low lighting and far-red (almost infrared) light illumination. Essentially, the Medusa allows scientists to spy on animals without having to sit in a cramped underwater vehicle.

In 2004, scientists using the Medusa were able to capture the first photo of a live giant squid in its natural habitat. In June 2012, scientists added an electronic jellyfish lure to the Medusa because the jellyfish's light is known to attract the attention of giant squid. They set off to a part of the ocean where sperm whales were present with the idea that, since sperm whales eat giant squid, there must be giant squid where sperm whales are located, and dropped the Medusa into the ocean. At a depth of about 2,000 feet, a giant squid popped into frame. The team then used a submersible to dive down and catch a closer look. For the first time ever, scientists were able to record video of the giant squid in its natural habitat.

The Medusa has permitted scientists to study the giant squid like never before. In 2019, off the shores of the United States, scientists recorded video of a giant squid stalking its prey at a depth of about 3,000 feet. Previously, it was believed that, given the giant squid's two long tentacles, it ambushed prey. Without this glimpse of the giant squid in its habitat, scientists would have never known of this behavior. And without the invention of a stealth camera system like the Medusa, we wouldn't have the tool we need to find the Kraken.

KRAKEN
SEARCH PROTOCOL

Our Kraken search protocol is based on the hypothesis that the Kraken is a giant squid living in the depths of the ocean. So, to find it, we'll utilize the same technology scientists use to study the giant squid.

1. Obtain a stealth camera system like the Medusa.

2. Attach e-jelly to the stealth camera system.

3. Locate sperm whales.

4. Deploy the stealth camera system more than 3,000 feet deep into the ocean and then at increasing depths.

5. Catch the Kraken on video!

OTHER OCTOPUS OR SQUID-LIKE CRYPTIDS

AKKOROKAMUI

This octopus or squid-like creature is said to be bright, almost incandescent red. It lives off the coast of the northern Japanese island of Hokkaido. Estimated at up to 360 feet long, it was sighted mostly in ancient times, though in 1980, cruise ship passengers in this area reported seeing an 80-foot bright red creature with tentacles.

LUSCA

Described sometimes as a half-octopus/half-shark creature and other times as a half-squid/half-eel creature, Lusca is said to roam the Caribbean Sea. It supposedly lives in blue holes, which are narrow pits in the ocean that can be 200 feet deep.

OKLAHOMA OCTOPUS

Deep within four man-made lakes located in Oklahoma, an octopus allegedly lurks! Octopuses are not freshwater animals, so exactly what people are seeing is unclear. But witnesses have reported a horse-sized creature with reddish skin and long tentacles. This cryptid has been blamed for unexplained drownings and the disappearances of swimmers, particularly in Lake Tenkiller and Lake Thunderbird.

ALTAMAHA-HA

Of all the cryptids in the world, lake monsters might be the most famous. Chances are you heard of the Loch Ness monster before you even knew what a cryptid was. But the Loch Ness monster isn't the only mysterious long-necked, serpentlike creature lurking in a lake. People have spotted similar beasts in lakes all over the world.

The Altamaha-ha, or "Altie," as it's become known, is a bit different. While it shares several features of a typical lake monster, this cryptid lives in a river instead of a lake. Now, you might be thinking that makes it easier to find. After all, rivers aren't as wide or deep as lakes, so how hard can it be?

Once you know a little more about the river this cryptid calls home, you'll understand the problem.

THE ALTAMAHA RIVER

Located in the United States, the Altamaha River stretches more than 137 miles. Starting in central Georgia, where the Oconee River and Ocmulgee River meet, it winds its way down southeastward, where it spills into the Atlantic Ocean. During this stretch, the river narrows and widens, curves and straightens, and splits and rejoins, all while also intersecting with other waterways. When it eventually nears the Atlantic Coast, it channels around islands in an area consisting of marshes, canals, and ponds. Notably, there are no dams along the Altamaha's main stem.

See the problem? With so many pathways to roam, the Altamaha-ha isn't confined to a single body of water like most lake cryptids.

THE LEGEND

Over the years, people have reported seeing Altie in different places along the Altamaha's mammoth river system. Often, these sightings are from a distance, but on at least one occasion, witnesses got an up-close and personal experience with this cryptid.

In May 1998, three 11-year-old boys were playing near a dock in the Altamaha River. No sooner had one of the boys jumped into the water than *something* gray and brown covered in seaweed or grass emerged from the water about 10 feet away. The boy's friends saw it first and began screaming for him to get out as quickly as possible. The boy turned, and upon spotting a large scaly tail, rushed up onto the dock and scrambled to safety.

Later, in reporting the incident to their mothers, the boys learned of Altie and the legend of the Altamaha-ha. To them, it fit. Positive they hadn't seen an alligator or manatee, they insisted they'd never seen any animal with a tail like the one belonging to the creature in the water that day.

OVER 100 YEARS OF ALTIE ENCOUNTERS

The boys' description of Altie matches descriptions from other encounters. Generally, people have reported Altie as a long, snakelike animal with a pronounced tail.

1830s ACCOUNT FROM CAPTAIN AND HIS SAILORS: 70-foot creature with the head of an alligator and a circumference about the size of a barrel

1920s–1950s REPORTS FROM TIMBERMEN, HUNTERS, AND BOY SCOUTS: large and snakelike

1960s SIGHTING FROM TWO BROTHERS: 20-foot dark beast with an alligator snout, horizontal tail, and triangular ridge atop its body

1970s ENCOUNTER: 5- to 20-foot creature with a snakelike head

1981 ACCOUNT: 15 to 20 feet long, snakelike with two brown humps about 5 feet apart that protruded from the water

1983 REPORT: 20- to 25-foot creature

1997 SIGHTING: 10- to 12-foot animal with three humps

2002 ENCOUNTER: over 20 feet long

2010 SIGHTING: something strange/odd shape

While it's good to have a consistent description of this cryptid, what's not helpful is how many different locations Altie has been spotted in over 150 years. Up and down the river, from where the Altamaha River opens into the Atlantic Ocean to creeks and rivers that meet up with the river inland, Altie is a wanderer.

For our mission to be successful, narrowing down where Altie may be hiding is going to be crucial.

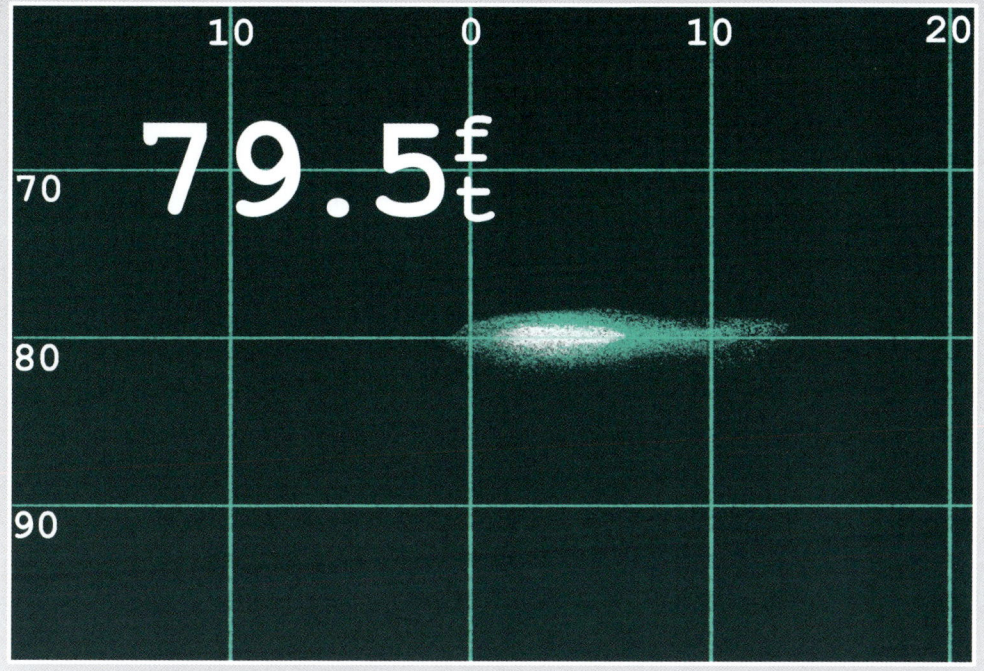

SONAR

Sonar (sound navigation and ranging) is a method used to detect underwater objects, such as submarines, shipwrecks, schools of fish, and other aquatic animals. It's been around for hundreds of years, but the early 1900s (right around World War I) is when it really caught on, as militaries developed sonar strategies to locate submarines.

Sonar navigation is possible because as sound travels through water, it bounces off objects and creates an echo. A boat using *active* sonar transmits sound pulses into the water, and, depending on how long each pulse takes to return, the transmitters are able to locate the object's position. If an object is nearby, it won't take long to hear the returning echoes. If the object is farther away, it will take longer. Essentially, sonar warns boats of potential dangers/objects in or around its path.

Many scientists use *passive* sonar, which is where sound signals are not intentionally transmitted into the water. Instead, passive sonar involves listening to sound waves already present that are being transmitted by aquatic life or other boats. If multiple listening devices are used, passive sonar can also measure the depth and distance of objects.

Sonar has been used to try and find lake monsters. In some cases, sonar actually detected large objects in lakes where people reported witnessing strange creatures. But sonar images can't identify or confirm the existence of an animal, much less a specific species. Their low resolution shows only the basic size and shape of the object that sound waves detected. In other words, nonbelievers can say the sonar image doesn't necessarily confirm the existence *of a large animal* versus vegetation, a school of fish, debris, logs, or some other big fish such as a sturgeon.

Sonar technology continues to advance, and with those improvements comes higher-resolution images. For now, sonar can help us by locating large objects in the Altamaha River that could be Altie so that we can continue our investigation with other technology.

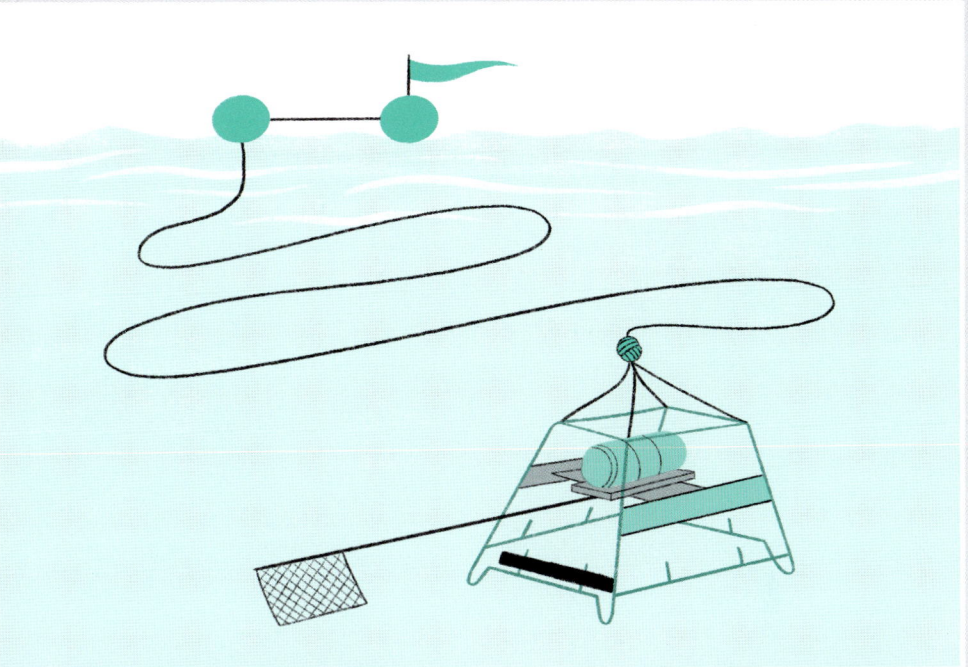

BRUVS

Baited remote underwater video systems (BRUVS) are small, light camera systems that utilize bait to attract fish to a high-definition camera that then snaps a photo. Scientists use them a lot in coral reef environments because the station can stabilize itself on rocky bases. One of the advantages of BRUVS is that they are motionless, unlike an intruding human diver who may scare off aquatic life.

BRUVS are routinely used by scientists looking at fish activity patterns, composition of a habitat, and the distribution of different fish in an ecosystem. In a study of angel sharks, scientists baited their BRUVS with mackerel and deployed six systems about 650 feet apart. In only 100 hours of coverage, the camera captured images of more than 48 species—including some that divers had failed to see—but no angel sharks. By obtaining video quickly through BRUVS, the scientists knew not to spend more time in that area and could search elsewhere.

Like all tech, BRUVS have their limitations. For example, in the angel shark study, some BRUVS cameras failed or were dragged by currents, and in other instances large spider crabs blocked a camera's view. Also, BRUVS are stationary in the water, so unless we deploy hundreds of BRUVS all along the 137-mile Altamaha River, it's possible for Altie to avoid being photographed altogether.

So, while BRUVS, like sonar, can certainly be helpful in locating this cryptid, our quest could use some additional tech.

ONBOARD CAMERAS

Onboard cameras rest or sit "onboard" an animal and allow scientists to learn about an ecosystem from the animal's eyes. Some cameras are so small that they fit on insects. Picture a camera about the size of a penny that weighs one-tenth of a playing card. Beetles wearing such cameras climbed rocks and flew without difficulty (showing that the added weight didn't affect their maneuverability).

Scientists using onboard cameras with tiger sharks, loggerhead turtles, and gannets (large, white seabirds) have learned things they never would have known through other monitoring methods.

TIGER SHARKS: Onboard cameras showed tiger sharks winding their way through tight coral corridors at speeds divers wouldn't have been able to match and at some depths scuba divers couldn't reach. Overall, the sharks' behavior suggested they used the coral to hide so they could sneak up on their prey.

LOGGERHEAD TURTLES: Scientists believed female loggerheads traveled alone and remained solitary during nesting season. However, the cameras showed female turtles interacting with other turtles.

GANNETS: Onboard cameras showed that gannets rely on other marine animals to direct them toward food sources. As soon as dolphins came to the ocean surface, gannets dove their way. Similarly, after a whale spouted water from a blowhole, gannets flew toward the whale.

Important for our mission is that these studies used onboard cameras in an *underwater* environment, meaning we now possess the necessary tech to implement a plan to find Altie.

ALTAMAHA-HA
SEARCH PROTOCOL

Our Altamaha-ha search protocol is based on the hypothesis that Altie is a large, unidentifiable aquatic animal in the Altamaha River. So, to find him, we'll use innovative tech that scientists utilize to study underwater animals.

1. Utilize sonar over 137-mile Altamaha River to locate large masses that may be Altie.

2. Place BRUVS in potential Altie locations as indicated by sonar.

3. View BRUVS camera footage with hopes of spotting Altie.

4. If BRUVS photos do not show Altie, choose an aquatic animal seen in the area of sonar-detected large masses for fit of onboard camera.

5. View footage from the onboard camera to locate Altie!

OTHER LAKE MONSTER CRYPTIDS

There are too many lake monsters to include in this book, but here is a list of some other well-known mystery lake creatures.

BESSIE

Bessie is Lake Erie's version of the Loch Ness monster. Since 1793, witnesses have reported spotting this approximately 30-foot, dark greenish brown or black long-necked sea serpent. In some circles, this cryptid is also known as Lake Erie Larry.

CADDY

Caddy makes its home in Cadboro Bay in British Columbia and has been seen for hundreds of years. In fact, it received its name way back in 1733. While descriptions vary, reports usually refer to a long-necked creature with a large head that resembles a camel or horse. It has a humped back or coils in the water and, depending on the witness, is anywhere from 16 to 100 feet long and may have both front and back flippers. Interestingly, more than one report has referred to the presence of *two* "Caddy" creatures seen together.

CHAMP

Lake Champlain, which borders New York and Vermont, supposedly contains this aquatic beast, sightings of which go back more than 400 years. Some describe Champ as being more than 180 feet long with a flat head and white star on its forehead. Other estimates are more conservative, placing Champ at 10 to 75 feet long. This lake monster is dark in color and displays humps or coils (between one and ten humps have been reported) while swimming with its head above water.

FLATTIE

Montana's lake monster resides in Flathead Lake, and it's another cryptid that's been seen for hundreds of years. It's been described as a whale-sized creature, serpentine in shape, between 20 and 40 feet long. Other reports say it's a giant fish 10 to 15 feet long, leading some to think there might be two unknown creatures lurking in this body of water.

IGOPOGO

We head north to Canada and Ontario's Lake Simcoe for this cryptid. Skeptics have said this aquatic creature is probably a seal, but witnesses describe it as 12 to 70 feet long with a dog- or horselike face and with multiple fins and a long neck like the Loch Ness monster.

KIPSY (THE HUDSON RIVER MONSTER)

This cryptid resides in the Hudson River in New York, where it has been seen between Poughkeepsie (which is where it got its name) and Manhattan. It's brown or gray and has been said to resemble a manatee (though one witness indicated the creature was 40 feet long, which would be one big manatee!). First reported in 1899, sightings increased between 2006 and 2009.

LIZZIE

Can you believe there are two Lizzies? One inhabits Lac Decaire, Quebec, and the other resides in Scotland. Lizzie in Scotland has gotten more attention due to 1937 footage that either depicts the beast . . . or doesn't. Witnesses in 1960 described Scotland's Lizzie as 30 to 40 feet long with a fin on the side of its body. Sightings in 1975 and 1996 described a 20-foot and 12-foot creature with three humps, respectively. Not much is known about the Canadian Lizzie, although her cousin "Memphre" might live in Lake Memphremagog, also in Quebec. Memphre is such a legend that it actually appears on a collectible coin released by the Royal Canadian Mint.

LOCH NESS MONSTER ("NESSIE")

Good ol' Nessie—the one that started it all! For whatever reason, this Scotland cryptid has captured and continues to capture the most attention of any lake monster. In some way, it makes sense, as there are reports of sightings back in 565 and 580 A.D. Numerous purported photos of Nessie exist, and sonar has even been used in an attempt to find her. The sonar results have been mixed, sometimes showing large unknown objects and on other occasions yielding nothing.

Over 23 miles long and nearly 800 feet deep, Loch Ness contains more fresh water than all the lakes in England and Wales combined. It even has its own small island and connects to other waterways. In other words, the loch offers plenty of room for Nessie to hide from people looking for her.

NAHUELITO

What secret creature can be found in Nahuel Huapi Lake in the mountains of Argentina? Why, it's our friend Nahuelito ("little Nahuel"), first observed in 1817 and repeatedly seen since. Estimated at 15 to 20 feet with a long neck, large body, and multiple humps, Nahuelito resembles the typical lake monster. Sightings continued into the 1990s.

OGOPOGO

Said to be a "cousin" of lake monster Igopogo, this cryptid also resides in British Columbia, specifically Okanagan Lake. Witness accounts vary, estimating Ogopogo from 20 to 70 feet long with a 5- to 10-foot neck and a horselike head that may or may not have stubby horns. Humps or arches of the creature's body have been seen in the water.

PEPIE

Lake Pepin is bordered by Wisconsin on one side and Minnesota on the other, and it's where this creature, a 100-foot animal with a long neck, lives. Sonar images have hit on a 16- to 17-foot *something* in the water, but no one has been able to procure a physical sample.

SELMA

We head to Norway for this lake monster. Residing in Lake Seljordsvatnet, this cryptid has been described as a serpentine beast with the head and mane of a horse, a 50- to 65-foot-long black crocodile, and other humpbacked, long-necked animals. One to five humps have been reported, along with a pair of front flippers. Photographs have been inconclusive, and while sonar has hit on large, submerged objects (two 60-foot-long animals!), accompanying film of the water was too dark to make out what was underneath.

SHARLIE

With the first reports occurring in the 1920s, this cryptid is just a baby compared to creatures seen for hundreds of years. Residing in Payette Lake in Idaho, Sharlie received her name through a newspaper contest. An estimated 35 feet long with a head resembling a snub-nosed crocodile, Sharlie has a humped back, with some witnesses seeing three separate humps in the water.

TESSIE

Tessie of Lake Tahoe, California, is said to be a brown, humpbacked, 12- to 25-foot-long creature. Sightings increased in the mid-1980s, and photos taken then and more recently are inconclusive. This region is no stranger to unusually large fish. In 2017, an 8-inch giant goldfish weighing several pounds was found in the Tahoe Keys, a wetland area south of Tahoe! (A typical goldfish is only 1 to 2 inches long.)

BEAST OF BUSCO

Ooooh! This cryptid has the word *beast* in its title. I wonder what kind of dangerous, weird creature we've stumbled across.

Spoiler alert: it's a turtle.

A turtle?

Well, to be more precise, a *giant* turtle.

THE LEGEND

The Beast of Busco came onto the scene in 1898. Oscar Fulk, a farmer in Churubusco ("Busco"), Indiana, told his neighbors a huge turtle lived in the lake on his farm. He didn't complain of attacks or other tales of mischief—just that a giant turtle was, well, *there*.

Fast-forward 50 years to 1948. Mr. Fulk had sold his farm, and the new owner saw *something* in the lake. Others did, too. Like the fishermen who said a giant turtle had tried to grab their fishing poles. After the new owner's chickens started disappearing, he asked local authorities for help catching the beast.

The entire town got in on the action. Clydesdale horses were used to pull the beast out of the lake, but the chains broke and the turtle (named Oscar after Mr. Fulk) escaped. Hunters and divers reported to the scene, but they were no match for Oscar. Reportedly, Oscar evaded stakes, chicken wire, and dynamite and eluded divers attempting to walk across the bottom of the lake. He even ignored a 225-pound female sea turtle brought in to tempt him out of the water. (The town assumed Oscar was a "he" who would be interested in a female turtle.)

By this time, the search for Oscar had been widely publicized. Thousands of people flocked to the farm. Airplanes flew overhead. Everyone hoped to spot the beast. On a Thursday in October, they got their chance. Oscar reportedly appeared while attempting to catch a duck being used as bait. Then, like before, he disappeared into the murky water.

The owner finally had enough. He decided to drain the lake. With a sump pump, the approximate 7-acre lake (think the size of about seven football fields) was slowly reduced to a mere acre. At that point, the muddy lake bottom wore down the pump. A crane was brought in to dredge the last of the lake, but the owner became ill and was hospitalized with appendicitis. By the time he was able to reconvene the search efforts, rain had refilled the lake. The search for Oscar was over.

Witnesses described Oscar as more than 6 feet long and 5 feet wide. Some reported its shell was the size of a dining room table or the top of a car.

No reports exist showing Oscar was ever seen after the lake refilled. But a year later, in 1950, two men draining swamps near Hammond, Indiana, a city a little bit north and about 150 miles west of Churubusco, encountered an enormous turtle blocking their drain.

They described its head as being as big as a human's and its shell as large as a beer barrel.

Is it possible Oscar had found a new home?

GIANT TURTLES

The most interesting feature of Oscar—the detail that puts him squarely into the "cryptid" category of *could this strange animal be real?*—is its reported size.

In modern times, the largest turtle ever documented is a leatherback turtle measuring 9 feet long and weighing more than 2,000 pounds (the average weight of a walrus!). The typical size for a member of a leatherback species is 3 to 6 feet weighing 660 to 1,100 pounds. In other words, these are big animals.

Certainly, considering size alone, the leatherback possesses the same "dining room table" feature as Oscar. Unfortunately, leatherbacks are *sea* turtles. They swim in oceans and don't live in fresh water. As a result, while the size comparison fits, the habitat doesn't, and we'll have to keep looking.

SALT WATER VS. FRESH WATER

The difference between saltwater and freshwater animals is how their bodies regulate water and salts. Nearly every living thing needs salt in its body because salt regulates fluids and makes nerves and muscles work. Since ocean water is very salty, saltwater animals are constantly losing salt through their skin. In contrast, freshwater animals constantly gain salt through their gills and skin. In both situations, the animal's body is regulating the amount of salt so that the salt level remains at a consistent level and the animal can survive.

The Galápagos Islands off the west coast of South America are known for several subspecies of giant tortoises. These turtles can get as big as 5 feet long and weigh more than 500 pounds. Alas, the Galápagos tortoises are *land* turtles that spend their days basking in the sun and using muddy puddles only for a relaxing soak. Given Oscar lived in a lake, the habitat doesn't fit for this species to serve as a good comparison, either.

For *freshwater* turtles—turtles living in rivers and lakes like Oscar—the giant soft-shell turtle is considered the largest in the world. But these turtles measure around 3 feet long and 2 feet wide—so certainly not as big as Oscar's reported size. More important, though, is that this turtle lives only in Southeast Asia, making it an unlikely explanation for Oscar, too.

ALLIGATOR SNAPPING TURTLES

So, if Oscar's not a leatherback, Galápagos, or giant soft-shell turtle, what is he? Well, if we narrow our search to large freshwater turtles living in the United States, it's time we take a look at the "dinosaur of the turtle world."

Alligator snapping turtles range across the southern portion of the United States from Florida to Texas and as far north as Indiana. They live in rivers, lakes, and wetlands connected to the Gulf of Mexico via the Mississippi River and its tributaries (streams or other water channels that lead to the river). Except for egg-laying females, these turtles almost never come on land. They prefer muddy bottoms and tend to stay submerged for long periods of time—they can go 50 minutes before needing to surface for air. In fact, they're motionless for so long that algae actually grow on their shells.

Those shells are thickly armored with three rows of spiked plates that resemble the hide of an alligator, which is how the turtle got its

name and is known as "the dinosaur of the turtle world." The shells range in color from nearly black to olive green (though some of that green could be due to the thick layer of algae). An alligator snapping turtle has a thick and particularly long tail—sometimes as long as its body. It also has an extremely large head with extraordinary strong jaws capable of supplying 1,000 pounds of pressure in a bite. That force is enough to snap a broom handle and even bite off fingers!

Don't worry. There's no record of alligator snapping turtles attacking humans. Although they are primarily carnivorous and eat fish, other aquatic animals, and small mammals, they also consume vegetation.

When it comes to size, males are larger than females. Specimens weighing between 225 and 300 pounds with shells up to 31 inches long have been documented, although there is an unverified report back in 1937 of an alligator snapping turtle that weighed more than 400 pounds. A 147-pound alligator snapping turtle caught in the wild in 2020 had a head the size of a basketball, legs bigger than human calves, and a nearly 2½-foot shell, while an alligator snapping turtle weighing 165 pounds was more than 4 feet long from head to tail.

Scientists believe this species' lifespan is 70 to 100 years, though in 2016 an alligator snapping turtle living in the Newport Aquarium in Kentucky passed away at an estimated age of 150.

THE OSCAR COMPARISON

Oscar shared several features with the alligator snapping turtle. Oscar lived in a murky pond with a muddy bottom. He often stayed submerged, yet he was tempted by a duck, revealing his carnivorous nature. Indiana is also home to alligator snapping turtles, and given the lifespan of this species, it would have been entirely possible for Oscar to appear in 1898 and again in 1948. As for his reported size, based on the dimensions of the 150-pound alligator snapping turtles, a 400-pound turtle could have a shell more than 5 feet long with a body near 9 feet. That means we're definitely approaching dining-room-table size!

FRESHWATER MONSTERS

While no one has been able to verify freshwater turtles the reported size of Oscar, abnormally giant creatures of other species have been found in fresh water:

In 2005, a 9-foot **giant catfish** weighing 646 pounds was pulled out of the Mekong River in Thailand. On average, giant catfish fall in the 4- to 6-foot range.

In 2018, a nearly 6-foot, 140-pound **Chinese giant salamander** was documented. On average, a Chinese giant salamander, which is entirely aquatic, averages 3.8 feet in length.

In 2011, a nearly 8½-foot, 327-pound **alligator gar** was found in a lake in Mississippi. On average, alligator gars are 5 to 6 feet long.

In June 2022, a 13-foot **stingray** weighing 661 pounds was located in a lake in Cambodia. On average, this species of freshwater sting-ray is about 7 feet long.

Remember: The eyewitnesses never saw *all* of Oscar—they saw his head and part of his shell in a pool of murky water—enough to realize he was huge, but descriptions like "top of a car" and "dining room table" can be the result of their brains "filling in the blanks" versus a precise measurement.

If we pursue the theory that Oscar was an abnormally large alligator snapping turtle, the next question is, where did he go? Could he have traveled 150 miles so as to have been the giant turtle spotted in Hammond, Indiana, the next year?

It's certainly possible, especially given that Fulk Lake was reduced from 7 acres to 1. Habitat loss and overcrowding prompt freshwater turtles to leave their habitat, and they use connected waterways (even temporary waterways, such as after a rainstorm or snowmelt) to travel from one body of water to another.

That Oscar wasn't detected during such a journey isn't necessarily unusual. Because alligator snapping turtles spend most of their lives submerged, often at the bottom of a lake or riverbed, they are hard to detect. Scientists have called them "secretive reptiles" and have historically experienced a difficult time locating them.

But that's not much help to us. There have been no reports of Oscar sightings for more than 75 years. Assuming Oscar survived the draining of his lake, escaped, and is still alive, where would we even start to look?

The story of an entirely different cryptid turtle located on the other side of the world might just show us the way.

HOÀN KIẾM LAKE GOLDEN TURTLE

Stories of the Hoàn Kiếm golden turtle began in the fifteenth century near the border of China and Vietnam in Asia.

Legend has it that while fishing, the emperor of Vietnam was given a sword by a giant golden turtle in Hoàn Kiếm Lake. This turtle became a symbol of Vietnam's independence and longevity.

In the 500 years that followed, witnesses reported seeing a golden turtle from time to time in Hoàn Kiếm Lake, but, given the lack of verification and the legendary tale, the turtle was considered a cryptid. Then, in 1967, a 550-pound, approximately 7-foot turtle crawled out of the lake onto the shore. The turtle died almost immediately. Scientists preserved and studied it, and some estimated it was 500 years old—the perfect age for it to be the legendary golden turtle.

But the story doesn't end there.

All was quiet on Hoàn Kiếm Lake for another 30 years when, in 1998, cameras glimpsed *another* giant turtle. This one was 6 feet long and 4 feet wide and, although not golden, it had a yellow-speckled face. It died in 2016 at an estimated 100 years old.

Scientists have identified both Hoàn Kiếm Lake turtles as the Yangtze giant soft-shell turtle, the largest freshwater turtle in the world. Unfortunately, they are critically endangered. In 2016, scientists believed fewer than four turtles of this species remained in the world. In 2019, another turtle died, and in 2023, the last known female also perished.

In an attempt to save and revive the species, scientists set out to find more, reasoning that there could be swamps and streams with similar habitats where other giant softshells might be living.

The problem these scientists faced was similar to the one we face with the Beast of Busco—knowing where to look. People hadn't reported seeing this species anywhere in the wild, and countless streams and swamps and wetlands served as potential hiding places. It's not easy to find a turtle that likes to live in murky water along muddy bottoms of lakes and rivers. How could scientists locate them?

With the advancement of environmental DNA (or eDNA), scientists received the head start they needed.

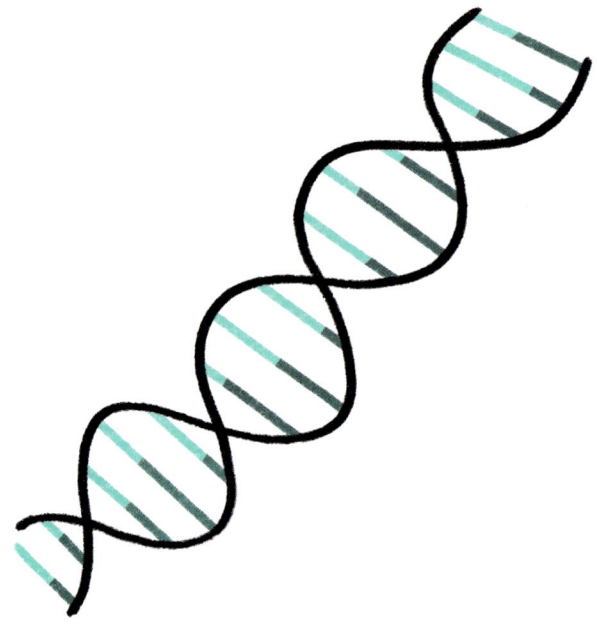

CONFIRMING NEW SPECIES—DNA

The primary way scientists identify an animal's species is through DNA. DNA stands for deoxyribonucleic acid, and it's in every cell of every living thing. Made of chemical substances linked together like a chain that forms a shape called a double helix, DNA has a very complex structure.

While every living thing's DNA is different, DNA from animals of the same species will share similarities. For example, if we looked at your DNA and my DNA, we'd have a lot of similarities, but there'd be about 3,000,000 differences, too. The similarities appear in how the chemical substances are sequenced in the chain.

So, if a scientist has a DNA sample from an unknown animal, it can look at its sequences and then compare it to known DNA sequences of other animals. Does the unknown animal's DNA appear similar to a wolf? A tiger? A bird? Is there a direct match, or is the DNA sequencing different enough to suggest a new species?

All of this is done using high-powered microscopes and other scientific instruments to extract and amplify the DNA, which is so tiny it's not even as big as the tip of a pencil. Scientists then use computer programs that compare the DNA the scientist fed into the computer with millions of sequences in its database to find similarities and potential matches.

ENVIRONMENTAL DNA (eDNA)

Environmental DNA (eDNA) is not DNA from a specific animal. Instead, it's DNA from a location; in other words, DNA from a specific *environment.*

The idea is that all animals shed DNA as they go about their day. Hairs, fibers, blood, feces, saliva, and so on all contain DNA. Therefore, a scoop of water from a stream should theoretically contain trace amounts of DNA from whatever animals recently passed through that area of the stream.

So, if scientists want to know if a certain species of turtle visited a stream, they take a sample of the stream's water to their lab and analyze all the different DNA in that sample. If DNA from the turtle is present, they know the turtle must be in the area *even if no one saw it.*

This is why eDNA is considered a breakthrough for locating hard-to-find animals. Now, instead of traipsing through wetlands, attempting to swim through muddy, murky water, or monitoring a river hoping to spot an animal, scientists can collect samples from different parts of a stream, lake, or river and learn what animals visited by analyzing the eDNA. It's a quicker, easier, and cheaper way to figure out if a secretive animal is nearby.

Through eDNA, scientists learned a giant soft-shell turtle was in Xuan Khanh Lake in Vietnam. This lake is about 12 miles from Dong Mo Lake, Vietnam, where one of the now-perished female turtles lived, and about 37 miles from Hoàn Kiếm Lake.

While eDNA has not yet located any other soft-shell turtles, the hope is that eDNA will lead to finding more, particularly a female who can successfully breed and revive the species before extinction.

Though alligator snapping turtles are not as endangered as the Yangtze giant soft-shell turtles, many states have implemented protections to prevent alligator snapping turtles from being hunted. Since scientists are already using eDNA to locate and protect alligator snappers, it's the perfect tool to use to start our search for Oscar.

BEAST OF BUSCO
SEARCH PROTOCOL

Our Beast of Busco search protocol is based on the hypothesis that Oscar is/was an alligator snapping turtle, so using science associated with studying that species will lead to finding Oscar.

1. Collect water samples from Lake Fulk and surrounding streams and lakes, including water bodies near Hammond, Indiana, where Oscar may have been seen in 1950, for environmental DNA analysis.

2. Analyze DNA in collected water samples for alligator snapping turtle DNA so as to determine if an alligator snapper recently visited the area.

3. Narrow physical search to bodies of water where alligator snapper DNA was detected.

4. Monitor bodies of water where alligator snapper DNA was detected. For example, in Vietnam, researchers use drones to monitor the lake containing the Yangtze turtle.

OTHER GIANT TURTLE CRYPTIDS

NDENDEKI (NDENDECKI)

This giant turtle lives near Lake Tele in the Republic of Congo, Africa. It has been described as 13 to 16 feet in diameter. The African soft-shell turtle is known in the area and can reach a length of 3 feet, though possibly as much as 6 feet.

PART TWO:

TAKING FLIGHT: WINGED CRYPTIDS

AHOOL

This cryptid's name is based on the sound it makes, a long *ahOOOooool* call, as it flies over your head.

  With its claws.

  And giant wings.

  Heads up!

THE LEGEND

The year is 1925, and Dr. Ernest Bartels is in a tropical rainforest on the island of Java. Java is in Indonesia, a country located in the Indian Ocean, northeast of Australia and south of Malaysia. The thirteenth-largest island in the world, Java contains nearly forty mountains, most of which were active volcanoes at one time.

The son of an ornithologist (a person who studies birds), Dr. Bartels is a naturalist who enjoys exploring the wild. While visiting a waterfall near Mount Salak, he suddenly views a large *something* flying overhead—a something that loudly screeches *ahOOOooool*. Unfortunately, it disappears soon after it's spotted.

But then . . .

Two years later, in 1927, Dr. Bartels is resting in his thatched hut, enjoying the sounds of the Java rainforest, when he again hears the familiar *ahOOOooool* cry. He isn't the only one, either. Sightings of what people in the area depict as a screeching giant bat have become routine. Witness descriptions are pretty consistent, too:

- Size of a 1-year-old child (about 26 to 30 inches tall)
- Wingspan of up to 12 feet
- Short, dark gray fur
- Flattened forearms supporting its wings
- Large black eyes
- Monkey-like head

During the day, the creature stayed in caves or behind waterfalls. At night, it patrolled the nearby river, using its large claws to catch fish. Finally, one witness reported seeing an Ahool upright with its wings folded before it darted into the sky.

Reports of Ahool sightings diminished over the years, but an estimated 225 species of bats live in Indonesia—the most of any country in the world. Should the Ahool be added to this list?

MICROBATS AND MEGABATS

There are two main species of bats: microbats and megabats. Microbats are—you guessed it—smaller in size. They generally have large ears and tails and range between 3 and 16 inches in length with wingspans reaching about 10 inches.

Megabats are the larger species. They have large eyes, lack a tail, and live only in tropical regions, like Indonesia. Wingspans commonly reach more than 3 feet.

Looking at size and habitat alone, it's easy to put the Ahool in the megabat category. By all accounts, the Ahool is a monstrous-sized bat, and it lives in a tropical rainforest.

Case closed.

Except that, other than size, the Ahool possesses quite a few microbat qualities:

- ☑ **FACE**: The Ahool's reported "monkey-like head" is consistent with microbats, which have more of a flat monkey face. Megabats' heads are more fox- or horselike.

- ☑ **POSTURE**: The Ahool's reported upright posture is consistent with some microbats that stand upright. Megabats do not stand upright.

- ☑ **DIET**: The Ahool's appetite for fish is consistent with microbats, which eat fish. Also, the only bat known to catch fish with its claws is a microbat. Meanwhile, megabats eat fruit.

It's because of these similarities with microbats that more than one cryptozoologist has theorized the Ahool, despite its mega size, may actually be an unknown species of microbat instead of a megabat.

Now, you may be thinking, "Who cares? Microbat or megabat—this bat beast sounds awesome. Let's go find it!" I hear you; I do. But bats are elusive creatures. They are predominantly active only at night, when they're difficult to spot, and they spend their days hanging out— literally—in hard-to-access trees or dark caves. Some species even fly above the forest canopy (tops of trees) and are highly maneuverable, making them difficult to catch.

Is there technology that can help our search? Absolutely. But that technology depends *entirely* on whether the Ahool is a microbat or megabat, so before we run off searching, let's take a closer look at both types.

MICROBATS

While in flight, microbats are continuously making noise through chirps, squeaks, and calls. They actually *have* to be noisy. Their vision is extremely poor, and it's only through these chirps and squeaks that they are able to navigate through the air. This process is called *echolocation.*

Echolocation is the ability to locate objects based on how those objects reflect sound (like sonar, as discussed in the Altamaha-ha chapter). During flight, bats emit chirps and/or squeaks and then listen for echoes. If an echo is louder and bounces back sooner, the bat knows the object is close. Echoes that take more time to return and are softer signify an object farther away. A sharp echo indicates a hard object (like a building), while a softer echo signifies softer objects (like tree leaves). By echolocating, bats are able to navigate through total darkness.

Since we know microbats are continuously making noises while in flight, one way to track and locate them is by listening to their sounds. The problem is the squeaks and chirps are usually pitched at too high a frequency for humans to hear.

Think of a dog whistle. Dogs are able to hear at a much higher frequency than humans, which is why dog trainers use high-pitched whistles—their dog hears the sound, but humans can't, and therefore we remain unbothered by the whistle's "noise." Bats communicate at an even *higher* frequency than dogs. Some bats' chirps are at frequencies ten times higher than the level at which humans can hear. Fortunately, bat detectors solve this problem.

DO YOU HEAR WHAT I HEAR?

Bats and dogs aren't the only animals that can hear and/or communicate at different frequencies from humans. These frequencies are measured in kilohertz (thousand cycles per second).

Here are some high-end ranges:

MOST FISH, AMPHIBIANS, AND REPTILES: 5 kHz
ELEPHANTS: 10 kHz
BIRDS: 8–12 kHz
HUMANS: 20 kHz
RACCOONS: 40 kHz
DOGS: 45 kHz
CATS: 64 kHz
MICE: 100 kHz
BATS: up to 212 kHz (depending on species!)

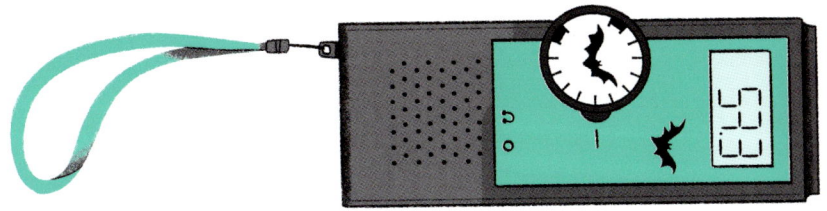

BAT DETECTORS

A bat detector is a device—many are about the size of a smartphone—that records the bats' high-frequency calls and then translates those calls into sounds humans can hear. Scientists use bat detectors to determine if bats are in the area.

Although there are approximately 1,400 bat species in the world, each bat species emits calls with specific characteristics related to their size, flight behavior, and environment. Advanced bat detectors are capable of recording these different calls continuously at several frequencies. The recordings can then be analyzed with software that identifies the specific bat species to which each call belongs. As a result of this sound analysis, scientists can learn which bat species are present in a specific area.

For example, in 2019, bat monitoring stations on Santa Catalina Island (off the coast of California) picked up signals of red bats—a species never seen on the island. From there, scientists were able to focus their research on when and how the red bat showed up as well as the role it played in the island's ecosystem.

Similarly, in 2022, scientists "rediscovered" the Hill's horseshoe bat—a species previously believed to have gone extinct after not having been seen for more than 40 years. The study began in 2019, when scientists captured an unusual-looking bat. They took measurements and recorded the bat's echolocation calls before releasing it back into the Central Africa rainforests. Bat detectors and monitoring stations throughout the rainforest then picked up recordings of the bat elsewhere, proving that, although it is a rare species, it is not yet extinct.

If the Ahool is a rare XXXXL species of microbat, bat detectors will pick up its unique echolocation calls. The detector will label them as "species unknown" since an Ahool cry is not on file/hasn't ever been recorded—and we'll know where to start our field research.

But, if the Ahool is a megabat, bat detectors are useless because most megabats *can't* echolocate (and the few who have a smidge of the ability can't do it effectively). Megabats aren't chirping or squeaking during flight, and they rely on their vision to navigate. The most audible sound we'll hear from megabats is the thumping of their large wings.

That means we're going to need different tech.

MEGABATS

A bat called the flying fox is the largest megabat. There are more than 65 different species of flying foxes, with the largest having a wingspan nearing 6 feet and a torso more than 1 foot tall. This is smaller than the Ahool (reportedly up to 12-foot wingspan and 1½ to 2 feet tall) but more in line size-wise than microbats. Plus, flying foxes can be gray—like the Ahool—and have large eyes.

Several species of flying fox are threatened or endangered; therefore, scientists have instituted methods to monitor and protect these bats from extinction.

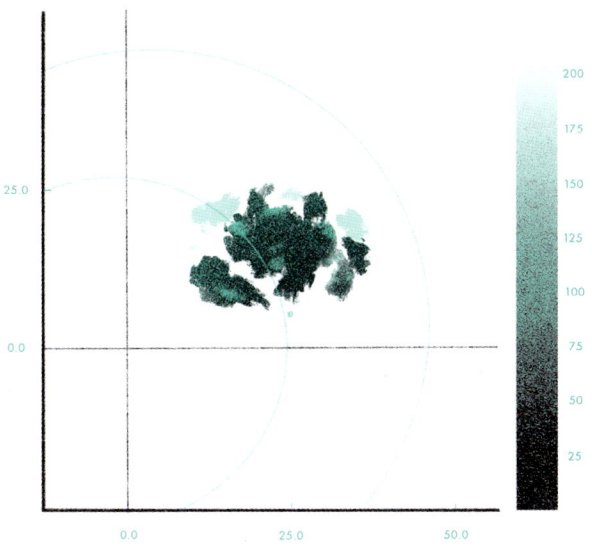

WEATHER RADAR

One piece of technology proven particularly helpful in locating and counting flying fox populations is weather radar. Flying foxes live in colonies of thousands. There could be upward of 50,000 bats in a single colony, while 1,000 bats are considered a small colony.

Because bats are nocturnal, as the sun sets, they take to the sky. Since thousands of flying foxes are skybound at the same time from the same roost, these "fly-outs" appear on weather radar systems in the form of tiny dots. By analyzing weather radar, scientists can estimate the size of a colony, the direction bats fly, and the colony's location.

However, weather radar has its limitations. It cannot detect the type of bat species present or track individual bats. In other words, we're not going to be able to pick out the Ahool from 50,000 flying foxes by viewing dots on a radar screen. But we will have colony location information, which means we can narrow down *where* in the Java rainforest to send our field research team.

THE JAVA RAINFOREST

The Java rainforest is much smaller today than it was in the 1920s, when the Ahool was first sighted. In fact, more than half of Indonesia's rainforests have disappeared since World War I due to deforestation from human activity.

When rainforests are destroyed, animals are displaced from their habitats and have to find new places to live. As those areas become more crowded with new and different species, animals already calling that habitat their home may find themselves displaced. When those animals then leave in search of a new home, the cycle repeats. Plus, as deforestation continues, the places available for an animal's new home decrease. Eventually, there's not enough room for everyone and certain species become extinct.

Java has already seen the effects of deforestation. The Javan tiger became extinct by the 1980s, and other species, like the Javan hawk-eagle, Javan gibbon, Javan slow loris, and Javan langur are fighting for survival. The Javan rhino has reached critically endangered status, down to only 40 to 60 individuals. The Javan mastiff bat, which is found only in western Java, is so elusive that scientists do not know how many remain.

Deforestation has occurred in the area where the Ahool was initially seen. What was once rainforest now contains a car dealership, soccer park, and shopping center. Did the Ahool find a new habitat deeper in the forest?

We can certainly place bat detectors and utilize weather radar to locate potential Ahool locations in case it managed to survive. At the same time, however, since bats are present on nearly every continent, I wonder whether there's somewhere else a creature like the Ahool might exist.

THE RED LIST

We often hear about "endangered" or "threatened" species, but what does that mean? The International Union for Conservation of Nature (IUCN) uses a "Red List" to identify species' extinction risk. Classifications include:

EXTINCT (EX)—beyond reasonable doubt that the species no longer exists

EXTINCT IN THE WILD (EW)—survives only in captivity and/or outside native range

CRITICALLY ENDANGERED (CR)—in a particularly and extremely critical state

ENDANGERED (EN)—very high risk of extinction in the wild

VULNERABLE (VU)—high risk of unnatural extinction without further human intervention

NEAR THREATENED (NT)—close to being endangered in the near future

LEAST CONCERN (LC)—unlikely to become endangered or extinct in the near future

THREATENED—include those falling under the Critically Endangered, Endangered, and Vulnerable categories.

The idea behind the Red List is that once governments know about a species' extinction risk, they will implement conservation efforts for that species.

SPECIES DISTRIBUTION MODELING (SDM)

Conservation scientists use species distribution models (think computer programs) to identify where species currently exist to determine potential suitable habitats for that species in the future. Knowing where else a species may thrive helps scientists protect that species in case its environment becomes threatened.

Scientists start an SDM by considering the entire world as an animal's potential habitat. Then, as information about a specific animal's environment is collected and entered into the system, the computer finds other areas on the planet where those same features exist.

The advancement of computer technology, including the ability to analyze large amounts of data, has led SDMs to become incredibly effective in identifying potential habitats for specific species. This is because, as is usually the case, the more information you have, the more accurate your estimation becomes.

Think of it like this: You're meeting a friend at a theme park. He tells you he's waiting in line at a ride, but then he stops texting. Now, you have to find him, so you weave through the crowd, checking attractions, trying to spot him. Maybe you'll eventually find him, maybe you won't. Without more information, you'll likely waste time looking in areas where he *never even was*.

Now, what happens if you have the added detail that he arrived at the entrance a couple minutes before you did? This information allows you to conclude that it's more likely he's closer to the entrance than at the opposite end of the park. With just this teensy bit of additional information, you've cut down the area you're searching by half.

Now, add this detail: he's afraid of roller coasters.

You can cross off roller coasters as places to look in the front half of the park.

Now include this detail: he's with his four-year-old brother.

Awesome! The children's section is right off the entrance.

With each piece of additional information, you've been able to narrow down where to look from an entire theme park to one section.

So, turning to our quest to find the elusive Ahool, we would give the computer information about the environment in Java where the Ahool was first spotted. The area's size, temperature, rainfall, species of trees/type of forest, soil composition, and elevation are a start, but it doesn't stop there. What bodies of water are nearby? Streams or lakes? What's that water's temperature? What species live in it? Don't forget the Ahool's diet. What kinds of fruit, insects, and fish populate the area? How tall are the trees for roosting? Are there caves nearby? How dense is the forest? How close is the forest to the island's mountains and volcanoes? What other species live in the same area?

All of these variables can be fed into a computer so that it knows as much as possible about the environment we're trying to match. Even better would be if we could use historical data to provide information on what the Ahool's environment was like in 1925, when it was first spotted. Then we put the computer to work to locate other regions where, based on the environmental characteristics we've provided, the Ahool may exist.

AHOOL
SEARCH PROTOCOL

Our Ahool search protocol is based on the hypothesis that the Ahool is/ was a large bat living in the tropical rainforest on the island of Java, so using tech associated with studying bats will improve our search efforts.

1. Analyze weather radar for information on the location of mega-bat colonies in Java.

2. In case the Ahool is a microbat, also place bat detectors/audio monitoring stations in Java where Ahool was initially spotted/ current sightings.

3. Analyze bat detector recordings for new/unidentified microbat calls that may belong to the Ahool.

4. Conduct field research at potential Ahool locations based on weather radar (megabat) and bat detector (microbat) results.

5. Have we found the Ahool? If not, gather data regarding the Ahool's environment when first seen in Java.

6. Use species distribution modeling to locate potential present-day habitats that match the Ahool's habitat in 1920s Java.

7. Repeat steps 1 through 4 at locations that SDMs indicated are a match for the Ahool.

OTHER GIANT BAT-LIKE CRYPTIDS

CAMAZOTZ

We travel to Central and South America for this giant bat. It reportedly has a bat-like head with a leaflike protuberance on the nose. One theory is that it is a species of spearnose bat, which has large nose-leaves and up to a 3-foot wingspan.

FANGALABOLO

This giant bat lives in Madagascar, an island country off the coast of East Africa. Although the megabat flying fox inhabits this island, reports indicate this cryptid's wingspan as larger than 5 feet, which is bigger than the flying fox's wingspan.

GUIAFAIRO

We stay on the continent of Africa for this cryptid, but we're traveling northeast. This giant bat has been spotted in Senegal, a country located on the coast of the Atlantic Ocean. It's been described as large and gray with clawed feet. Witnesses also mention a nauseating/foul odor when it is around.

KONGAMATO

This cryptid was first reported in the 1920s around the same time as the Ahool. Seen near Zambia (a country in the central southeastern part of Africa), this cryptid's torso is reportedly 2 to 4 feet long. It is black or red in color and has a long, narrow tail, long beak, and teeth. Its wingspan is 3 to 7 feet. Additional sightings occurred in the 1950s. Some wonder whether this giant bat is the saddle-billed stork, which is 5 feet in length and has a 9-foot wingspan. Others posit it's related to the hammerhead bat—a megabat that lives in the region and has more than a 3-foot wingspan.

JERSEY DEVIL

Located in the Pine Woods of New Jersey, this cryptid has allegedly been encountered by numerous people since 1735. While descriptions of this creature's body and head vary from a horse to a goat to a dog, one common characteristic is its giant bat-like wings and long tail. Overall, it's between 3 and 11 feet long and more than 2 feet wide. As with the kongamato, some think a species of hammerhead bat may be the answer. The problem with this idea, though, is that the climate in the United States isn't right for megabats, which live in tropical regions, so it's unclear how an entire species could end up in New Jersey, much less enough specimens to repopulate over 300 years.

OLITIAU

The olitiau lives in central Africa and is often compared to the konga-mato. Sighted in 1932 by a cryptozoology team in southern Cameroon, it was described as being about the size of an eagle and having a black body, bat-like wings, a wingspan of 6 to 12 feet, and 2-inch-long teeth. The team reported their find to locals, who expressed familiarity with the "granddaddy of all bats" and then advised the cryptozoologists it'd be better to move their camp elsewhere. Additional attempts to learn more about this creature failed due to language difficulties, and the team left without understanding anything other than that the creature was believed to be a "bad" animal.

ROPEN

This cryptid inhabits Papua New Guinea, a country that comprises the eastern half of the island of New Guinea, which is located a smidge north of Australia. Described as a "large flying reptile" with a long neck, sharp jaws, long tail, and a 4- to 20-foot wingspan, it resembles a pterosaur or pterodactyl (i.e., a "flying dinosaur"). Witness reports from the 1990s and early 2000s have this creature swooping down from the sky and attacking boats and snatching fish off them.

MOTHMAN

It's not a moth. It's not a man. And yet, somehow, for some reason, this flying creature that witnesses have described as a "giant bird" goes by the name Mothman.

Go figure.

Mothman's name isn't the only strange thing about this cryptid. Unlike many other mystery animals, Mothman hasn't been seen in only one area. People have reported encounters for more than 60 years across several states and in locations as varied as the woods, along highways, and atop skyscrapers in large cities.

Is everyone seeing the same thing? If so, how are we ever going to track it down?

THE LEGEND

On November 15, 1966, two couples were driving down a highway in Point Pleasant, West Virginia, when they encountered a birdlike creature resembling the shape of a man with fiery red eyes that glowed in their headlights. The creature stood approximately 6 feet tall and had large wings. As they drove away, the winged beast flew into the sky, where it glided above the trees before returning and following them down the road. The creature was fast. In trying to escape, the couples sped 100 miles per hour, and the creature was able to keep pace before eventually disappearing into the night.

In the weeks and months that followed, others spotted a similar sized giant bird-thing with red eyes. Some were positive the creature they saw was a bird. Depending on the witness, the creature's wingspan was between 6 and 10 feet. There were even reports of two Mothmans hanging out together.

News of a large, winged creature terrorizing the area spread, and media dubbed the creature "Mothman." Despite searches of a nearby abandoned World War II munitions site and wooded area, no physical evidence of the creature was found.

As Mothman's reputation expanded, its legend grew. Some suggested it was a paranormal creature; others said its presence caused or predicted bad events, like a bridge collapse in 1967. The descriptions, though, remained somewhat consistent, even when reports started coming in from other states. Mothman was a large, winged creature with red eyes that sometimes screeched as it flew overhead.

MOTHMAN MAYHEM

Despite the frightening encounters over the years, Point Pleasant, West Virginia, has embraced Mothman's legacy. Every September, the town holds the Mothman Festival to commemorate the 1966 sighting. Thousands of visitors flock downtown to enjoy live music, food trucks, cosplay, and other attractions. The Mothman Museum, which is dedicated to all things Mothman, is also in the area.

PRESENT-DAY SIGHTINGS

In contrast to Mothman's beginnings in rural West Virginia, more recent sightings place the beast in Chicago, Illinois. In March 2017, a truck driver viewed a creature the size of a large car "gliding like a bird" in the sky. A month later, a witness encountered a 7-foot-tall "manlike" bird in a Chicagoland park.

Similar accounts of large flying creatures followed throughout the summer, with some seeing it dive off a skyscraper. At its core, the Chicagoland reports described a large gray, black, or brown bird-man, big owl, or birdlike creature with a 12-foot wingspan that glided through the air. Interestingly, the "red eyes" detail is absent from some of these 2017 reports.

Chicago Mothman sightings continued, and in 2020, a woman reported seeing a 7-foot red-eyed creature screech as it took off to the sky after striding toward her outside O'Hare International Airport. Similarly, in 2021, a different airport employee heard loud screeching and turned to find a large creature with red eyes staring at her. The beast was black with an estimated 10-foot wingspan and shot into the sky. Finally, in 2023, another person stationed at the airport encountered a 7-foot black creature with red eyes and a 15-foot wingspan before it flew into the air. Instead of a screeching sound, a few witnesses reported hearing rapid clicking or chirping.

So, where do we go from here? With varying descriptions and no precise place to look, it'd be easy to get overwhelmed trying to find Mothman. But since the majority of witnesses mentioned bird and birdlike features in their descriptions, let's see if there are any known birds out there that bear some resemblance to this cryptid.

BIG BIRDS

While Mothman's height and wingspan vary from account to account, it's safe to say we're looking for a winged creature around 6 feet tall with a large wingspan. Whether that wingspan is 10 feet, 12 feet, 15 feet, or something else (it's not like anyone had a tape measure out there), we can put that aside for now. Let's take a look at some birds with large wingspans.

SANDHILL CRANE

One theory is that people spotting Mothman actually caught an unexpected glimpse of a sandhill crane. Sandhill cranes are big birds: the average height is up to 4½ feet tall with a wingspan of more than 7 feet. They have a long neck, black legs, a bright red, fleshy area around each eye, and live across many northern and southeast states, breeding in the northern areas and then migrating south for winter. Attracted to large open spaces, sandhill cranes have been known to cause problems near airports—which is an interesting detail given how many Mothman encounters occurred at or near O'Hare outside Chicago.

Some witnesses mentioned Mothman was gliding versus flapping its wings like other birds. This ability to stay aloft for hours is a known feature of the sandhill crane. Their large wingspans allow them to use the air and soar with only occasional flapping of their wings.

As for Mothman's travels across several states, sandhill cranes can fly more than 400 miles a day. One of the oldest sandhill cranes ever monitored was born in Florida and died in Wisconsin over 37 years later. They also can reach speeds exceeding 50 miles per hour, which could certainly enable them to keep pace with a car speeding down a highway.

A sandhill crane could be responsible for both the screeching sound and the clicking noises witnesses reported. Due to their long windpipes, sandhill cranes are able to vary their calls. In fact, sandhill cranes have at least eighteen different vocalizations. The calls differ in intensity and volume depending on the crane's activity. Piercing rattles may signal a predator is near, while purring clicks occur before the crane takes flight.

Finally, let's get to those reports of glowing red eyes. Given the red area around the sandhill crane's eyes, car lights/headlights could cause the skin to reflect as big red circles around the eyes. This could also explain why some witnesses didn't report/mention red eyes— perhaps the area was not light enough or light was not directed at the sandhill crane's face to pick up the red area.

Hmm. Mothman certainly shares several characteristics with the sandhill crane with one exception—a very long neck—that none of the witnesses noted. Is that a detail people would leave out? It'd be like being terrorized by a giraffe and mentioning its four legs, tail, head, but *not* its neck.

Of course, it's possible shadows or the crane's positioning could make the neck seem shorter and not worth mentioning. Given these sightings occurred in the dark, if a crane is standing with its wings extended, a person might not be able to distinguish its neck.

But before we settle on the sandhill crane, let's see if any other birds are Mothman-y enough for a match.

Barn owl

Barred owl

Great horned owl

OWLS

When most people think of an owl, they probably think of a small animal perched on a branch *hoo-hooing* away in the dark night. Fair enough. But owls come in all shapes and sizes, and some possess enough similarities to Mothman's reported description that they deserve closer scrutiny.

Right from the start, an owl's overall shape is consistent with Mothman's. Owls don't have defined necks, so there'd be no reason for witnesses to specifically describe a "neck" in their reports. Owls are also known for red eyeshine. Eyeshine is the shining or glowing effect in an animal's eyes at night. It's common in nocturnal animals because animals active at night develop differently to help them see in the dark. When light hits their eyes and passes through the retina, it reflects back through the retina in *twice* the amount, which makes the eyes "glow." Glowing red eyes are commonly seen in owls as well as alligators, cats, rabbits, and seals.

WHAT BEAUTIFUL EYES YOU HAVE

Red isn't the only color associated with eyeshine. You may also encounter animals with yellow, green, blue, or white glowing eyes!

YELLOW: bears, cats, deer, raccoons, panthers, some owls
GREEN: some dogs, cats, foxes, opossums, badgers, sheep
BLUE: horses, woodchucks, some dogs
WHITE: coyotes, deer, tigers, many fish

One potential owl suspect for Mothman is the barn owl. **Barn owls** have smoothly round heads, rounded wings, and a loping flying style. They live in nearly every state, coast to coast, north to south, including West Virginia and Illinois. Barn owls like open land, whether grassland or fields, and roost in barns and abandoned buildings. Instead of hooting, this owl makes a long screech or scream sound.

While those features may be a good match for Mothman, barn owls fall a bit short on size. They can have wingspans over 4 feet, but their bodies are only about 16 inches tall.

Barred owls are another option. Barred owls are a bit larger than barn owls and their eyeshine is stronger—meaning they're a perfect culprit for bright glowing red eyes. They like living in forests, including the area in West Virginia where Mothman was first reported. Barred owls make more of a hooing sound than a screeching noise, though, and the length of their bodies is often less than 20 inches—meaning we're still well under that 6-foot height characteristic.

The **great horned owl** is taller than barn and barred owls. Its body can be more than 2 feet in length and it can have a 5-foot wingspan. It, too, has a rounded head and rounded wings, and while it enjoys the woods and open areas, it is also known to hang out in cities and suburbs. As for activity, the great horned owl is a hefty bird known for silently drifting through the night before pouncing on predators. Its call is a booming hoot as opposed to a screech or scream.

Sigh. Like the sandhill crane, certain owls appear to share some of Mothman's features but not others. That means our investigation continues.

TURKEY VULTURES

Turkey vultures are large, dark brown (often mistaken for black) birds almost 3 feet tall with wingspans near 6 feet. They have red heads (like turkeys) and live across North America, where they are commonly seen in fields and along roadsides, but also near food sources, such as landfills and construction sites. Like other birds with large wingspans, turkey vultures glide through the air without often beating their wings.

Even though they're not constantly flapping their wings, turkey vultures are fast—they can fly up to 60 miles per hour—and often soar through the sky before dive-bombing their prey. Like the sandhill crane, turkey vultures get around, too. They can travel up to 200 miles a day.

However, turkey vultures don't hoot or screech—they don't have vocal organs to make proper songs. Instead, they emit a low, guttural hiss or hissing grunt.

And turkey vultures are predominantly not active at night. They scavenge during the day and spend nights roosting. Because they're not nocturnal birds, they do not have glowing red eyeshine like owls, so any red seen would be from the red on their heads, much like the sandhill crane theory.

WHAT NEXT?

So, what can we make from all this? For one thing, after examining sandhill cranes, owls, and turkey vultures in conjunction with Mothman descriptions, it may very well be that everyone is not seeing the same Mothman. Mothman's vocal calls and facial features vary between witness accounts, as do the habitats where it was spotted.

This means our quest has changed from searching for Mothman to searching for Mothmans. Perhaps one is a sandhill crane while one is a turkey vulture while another is an owl. Or, maybe one of these Mothmans is a bird that has not yet been scientifically identified.

Perhaps, instead of focusing on what Mothman looks like, it's time for us to do some listening.

BIOACOUSTICS

Bioacoustics combines the fields of biology (science of living organisms) and acoustics (science of sound) by using audio recordings to detect, identify, and monitor species within a certain habitat. One of the best things about audio recorders is their ability to detect sounds from all sorts of different animals—birds, mammals, insects, etc.—from greater distances than a person can hear when standing in the same area.

Having all that data poses a problem, though, because of the amount of time it takes for someone to sift through hours upon hours of recordings to locate a specific animal's call. Some animals use low or high frequencies that are hard for the human ear to detect.

Also, audio recordings of a habitat record *everything*, whether it's a chirp, buzz, growl, or wind whipping through the trees. That makes it even more difficult for scientists to detect the sound they're interested in. Picture the forest as a 1,000-piece orchestra where you're trying to detect one specific note from a flute—it's not easy to pluck out that one note in a symphony of music.

That's why recent advancements in technology and artificial intelligence (AI) have been so significant. With the right software, algorithm, and program design, a computer can listen to recordings and pick out the sound the scientist is looking for.

For example, for nearly a hundred years, people heard strange squawks resonating from the forests of an uninhabited part of a small island off the western coast of Africa. In the early 1990s, a strange owl was spotted among the trees. Was this owl the source of the sound? To find out, scientists set out to catch it.

First, scientists deployed audio recorders in the forest. Then, they analyzed the recordings for the mysterious owl's call—a short, repeated squawk described as the mewling of a cat or rasp of an insect. Once scientists identified the call, they designed a computer program to sift through hours and hours of recordings to identify the bird's location.

That research led to snapping the first photo of the owl in July 2016, and then, in May 2017, an owl was finally caught. In 2022, after DNA analysis and additional studies, scientists finally confirmed its identity as a new species of a tiny, yellow-eyed owl.

Scientists researching the Hawai'i 'amakihi, a small yellow bird, also relied on bioacoustics. After recording the bird's call, they devised an algorithm for a computer to search for the sound from recordings taken in a forest. The call was found in wet lowlands—an area where the bird had never been detected before, leading scientists to a new location to study their bird.

Bioacoustics isn't perfect—there are still times when software will miss a sound or misidentify an animal. But it's better than pre-technology days, when scientists would trek into the forest and spend weeks taking notes on what sounds they *thought* they heard. Plus, audio recordings can be used over and over again by different scientists studying different animals. All they have to do is devise an algorithm to detect the specific animal's sounds they're looking for.

Birds are one of the most common animals studied using bioacoustics. Calls of sandhill cranes; barred, barn, and northern horn owls; and turkey vultures have already been collected and are available for comparison purposes.

Of course, hearing a call on a recording is only half the battle. After sound recordings alert us to what birds have visited an area, we'll need to take steps to capture visual evidence of their presence.

CAMERA TRAPS

A camera trap is a camera that is activated by movement. A camera with a sensor is attached to a tree or other structure, and, when an animal moves into the sensor's range, the sensor triggers the camera to take a photo. Camera traps are particularly useful for hard-to-detect animals. Instead of deploying a research team to walk through a dense forest looking for traces of animal activity like dung, footprints, or nests, scientists can set camera traps and then, after a certain period of time—days, weeks, etc.—review the photos to determine the area's animal residents.

For example, in 2015 in Gabon, a country in central Africa covered with dense forests and hard-to-access areas, camera traps detected honey badgers and aardvarks—two animals not previously seen in the area—and footage of a lion that hadn't been seen in 20 years. In 2019, camera traps captured photos of the Vietnam mouse-deer, which hadn't been seen in more than 30 years and was feared to have gone extinct.

Certainly, seeing an animal on camera verifies its presence in an area. But camera traps have their drawbacks. Unlike audio recorders, which can pick up sounds in the distance, an animal must be close enough to the camera's sensor and visible in the lens to be captured. Animals visiting the area can evade detection by being behind the camera or off to the side.

Since the camera has to be a certain distance above ground to take a useful photo, cameras have also been seen—and destroyed—by animals. Sometimes the camera makes a clicking noise, which may deter animals from getting close enough to trigger the sensor. If the camera is set to take color photos, it will also emit a flash, which may scare animals from revisiting the area. Infrared flashes are not visible, but they produce black-and-white photos only, which is fine if scientists merely want to detect the presence of a species but not as useful if scientists are using camera traps to monitor species population.

One of the biggest problems with camera traps is that researchers can end up with hundreds of thousands of photos. Depending on how many camera traps were deployed on a particular project, it could take scientists over a year to simply go through 1 year of footage for one study.

As in bioacoustics, though, technological advancements have been a tremendous help. There are now programs that will sift through all the images and remove blank images (for example, when wind, leaves, or something unseen triggered the sensor). In addition, computer scientists have worked with researchers to develop programs that allow a computer to review all the images collected in a study and mark which photos depict the species being researched.

Of course, to create a computer program that can accurately identify an animal from a camera trap photo, we need images of that animal. That's why we're in luck that there are many images of sandhill cranes; barred, barn, and northern horn owls; and turkey vultures available to create our database.

MOTHMAN
SEARCH PROTOCOL

Our Mothman search protocol is based on the hypothesis that Mothman is a large bird—but not the same large bird—sighted between Chicago and Point Pleasant, West Virginia, so using cutting-edge tech for studying birds will improve our search efforts.

1. Deploy audio recorders at recent Mothman sighting locations.

2. Create computer algorithms to analyze audio recorders for calls of Mothman suspects: sandhill crane; barred, barn, and northern horn owls; turkey vultures; and "unknown" species.

3. Identify the location of recorders where calls of Mothman suspects were obtained.

4. Install camera traps in and around locations where audio recorders picked up calls of Mothman suspects.

5. Use a computer program/AI to sift through all camera trap photos to find images of Mothman suspects.

6. Deploy a field team to the area of camera traps where Mothman suspects were photographed.

7. Find Mothman/Mothmans!

OTHER GIANT BIRDLIKE CRYPTIDS

SLAGUGGLA

This large owl lives in northern Europe and is said to have a wingspan of 10 feet and the ability to lift large animals into the air. The word *slaguggla* is Swedish for strike owl, though breeds of the strike owl have wingspans of about 2 feet.

BIGHOOT AND GIANT OWLS

Sightings of owls as tall as Mothman have been reported for hundreds of years. In 1982 and 1983, a witness reported seeing a 10-foot owl that later became known as Bighoot in Ohio. Some wondered if Bighoot could be Mothman.

In the 1970s, a creature dubbed Owlman and described as a big, furry bird with a gaping mouth and big eyes was seen in England.

THUNDERBIRDS/BIG BIRDS

Sightings of unidentified big birds are not uncommon, with encounters occurring throughout the United States. Witnesses have described black, gray, or brown birds 3 to 8 feet tall with wingspans of up to 30 feet.

A report from 1977 has one such large bird picking up a 10-year-old boy and carrying him 30 to 40 feet before dropping him due to his mother's screams. In the 2 weeks that followed, sightings of the same or similar bird were reported eight times, but no physical evidence was ever found.

PART THREE:

GROUNDED: LAND CRYPTIDS

MONGOLIAN DEATH WORM

This cryptid's name gets straight to the point. Reportedly, this creature lives in Mongolia, spits deadly venom, and resembles a worm. But even though it's been the subject of multiple crypto-zoological expeditions, physical evidence of its existence remains elusive.

Is this worm real? If so, why is it so hard to catch?

Perhaps the answer lies in its name. What if it's not a worm at all?

THE LEGEND

This cryptid was introduced to the world in 1926 by paleontologist Roy Chapman Andrews. He was in Mongolia, more specifically the Gobi

Desert, looking for dinosaur fossils, when the locals told him about "an intestine worm" living in the desolate sand dunes in the southern portion of the desert. He never personally saw the worm but included the warning in a book recounting his travels.

In the years since, this cryptid has been described as having a wormlike appearance with no obvious head or tail. It's as thick as a man's arm and ranges between 2 and 5 feet long. Some have stated it is dark red with darker spots or blotches, though others have indicated it is more brownish. It moves in a sidewinding motion, squirming sideways, similar to some desert snakes.

As for the "death" part of its name, locals described incidents where the worm rose up out of the sand, inflated itself, and squirted toxic venom at its victim. One report indicated a boy accidentally stumbled across the worm while camping with his family and died after merely touching it. Finally, some witnesses added that this cryptid also gives an electric shock.

WORM HUNT!

The Mongolian Death Worm has captured the attention of several scientists and cryptozoologists. In the 1990s, a team of researchers tried luring the worm out from underground through a series of vibrations, thinking the disturbance would compel the worm to the surface. It did not work. The same team tried again in a different region of the Gobi Desert a couple years later, this time using explosives to bait the worm aboveground. Again, they were unsuccessful. Their last attempt in 1996 also ended in failure.

A decade later, in 2005, a different cryptozoologist set out with his own team to find the worm. This team set up bucket sand traps strung together with mesh netting, the idea being that the worm would bump into the netting and then crawl along it until it fell into a bucket. By the expedition's end, they had not found the worm, either.

Given the massive size of the Gobi Desert, it's not totally surprising these expeditions weren't a success. Stretching from Mongolia's southern border with China to inner Mongolia and over 500,000 square miles, the Gobi Desert is nearly the size of California, Montana, and Texas combined. That's a lot of area to look in for what could be a 2-foot-long cryptid.

To ensure our mission doesn't meet the same fate, it would help to narrow *where* in the Gobi Desert we should be searching. And, perhaps most importantly, what exactly is it that we are looking for?

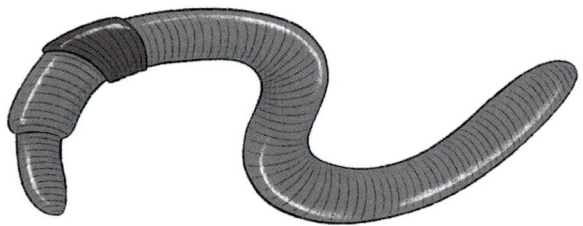

THE WORM THEORY

This cryptid's description as long, slender without limbs, and reddish in color certainly resembles a worm. That it's been described without visible eyes is also consistent with worms, as worms do not have eyes, ears, or a nose.

The Mongolian Death Worm's behavior is also consistent with earthworms, as earthworms emerge aboveground after rain. As for its reported 2- to 5-foot length, there are earthworms that can get that big. Giant beach earthworms in Australia live in the sand and can grow up to 10 feet long. This creature's desire to stay below the surface in the sand is also consistent with giant beach earthworms, which can burrow up to 5 feet deep.

There are a few differences, though. The beach earthworm is significantly thinner in circumference. It's also brown, not red. Finally, the giant beach worm makes a weird gurgling sound—something not reported by witnesses to the Mongolian Death Worm—and, perhaps most importantly, it doesn't spit venom.

The purported venom-spitting ability is something not seen in any species of earthworm. Combined with the other non-wormlike features this cryptid possesses, it's one of the reasons some believe this so-called worm may actually be a snake.

THE SNAKE THEORY

Many venomous snakes spray poison. Some spitting cobras are known to spit venom up to 10 feet. There are also numerous burrowing snakes with cylindrical bodies that live in the desert. As for the color, red venomous snakes, like some species of vipers, do exist.

But while the color, venom, and habitat are a match to the Mongolian Death Worm, the snake comparison falls short in overall appearance. Witnesses have described the worm as a segmented animal without eyes and lacking a defined head or tail.

Snakes have distinct heads. Vipers are known for their triangular head, and the puff adder—the most venomous snake on the planet—has distinctive banded coloring around its head and along its body. We also should keep in mind that the Gobi Desert is home to many snakes, so there is no reason to believe a local wouldn't refer to this creature as snakelike instead of wormlike if it indeed resembled one.

So, if it's not a worm or a snake, where does that lead us? Well, to the fabulous legless lizard, of course!

THE LEGLESS LIZARD THEORY

Legless lizards are cylindrical in shape and can resemble worms and snakes. They predominantly live underground, like worms, but move like snakes when on the surface. They have tiny eyelids and external ear openings and are able to blink, unlike snakes. They also possess a notched tongue instead of a snake's forked tongue.

Legless lizards are found in drier areas and, like earthworms, they often come to the surface after it rains. Some species are pink (like earthworms), while others are brown (like some snakes). Depending on the species, they measure anywhere from a few inches to a few feet.

Because they spend the majority of their lives underground, legless lizards can easily go undetected. In 2013, four new species of legless lizards were discovered in California, where researchers noted that the animal spends its entire life in an area the size of a small table. In other words, these aren't animals that roam vast distances. Finding one requires pinpoint accuracy to the animal's underground home.

That home could be anywhere on five continents. Most legless lizards are found in South America, but there are legless lizards in the general area of Mongolia. As is true with earthworms, however, there are no known venomous species. That being said, there are many species of venomous lizards, so some might say the potential for discovering an as-of-yet-undetected venomous legless lizard is still possible.

AND THE ANSWER IS . . .

So, what exactly is the Mongolian Death Worm? A worm? A snake? A legless lizard? Unfortunately, while locals in the Gobi Desert appeared to have encountered *something*, without additional information, we're unable to pinpoint exactly what that something is at this time. That means instead of devising our search protocol around a specific animal, we'll have to rely on technology that can locate this cryptid no matter what it turns out to be.

> ### WHAT'S IN A NAME?
>
> If the Mongolian Death Worm turns out to be a snake or legless lizard, it won't be the first creature to be burdened with an inaccurate name. Here are a few more:
>
> **Centipede** is Latin for 100 foot, but, in actuality, centipedes cannot have exactly 100 legs. They have only one set of legs per body segment, and their segments are always an odd number.
> **Electric eels** aren't actually eels. While they have an "eel-like" appearance, they're freshwater fish belonging to the knifefish family.
> **Horny toad** may look like a toad, but it's actually a lizard.
> **Starfish** aren't fish. They don't have gills, fins, or scales.

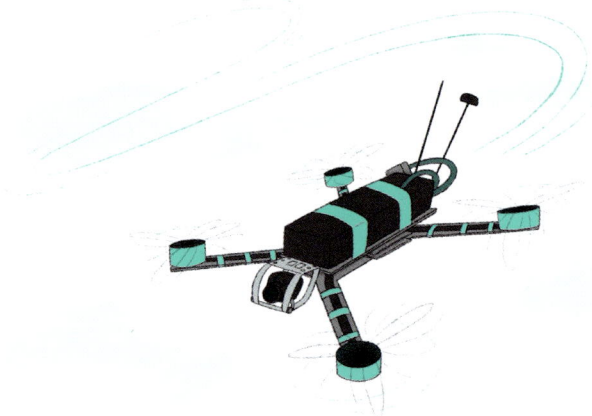

UNMANNED AERIAL VEHICLES (UAVs) AND DRONE MAPPING

Unmanned aerial vehicles (UAVs) are more commonly referred to as *drones*. Basically, a drone is a flying robot that can be controlled remotely or programmed to fly a specific flight path. With GPS, sensors, and the right software, drones can provide scientists with information that may otherwise be difficult or painstakingly slow to obtain.

For example, in 2020, the Wisconsin Department of Natural Resources (WDNR) used drones to treat phragmites. Phragmites is an invasive weed that outcompetes native plants and displaces native animals in wetland areas. Its presence indicates the ecosystem is out of whack, and, if uncontrolled, it can cause major problems, including fire risks.

The WDNR wanted to treat the phragmites with a spray that would kill it but not harm the surrounding area. To do that, it needed to accurately pinpoint where the weed existed. On foot, it would take 3 months for a person to map 2,000 acres. They would have to kayak the wetlands with a handheld GPS and carefully mark locations of the weed.

Using a drone, it took less than 3 days. The drone was outfitted with a high-resolution camera and appropriate software and flew a designated path over the area for 12 hours. Over 13,000 images were collected and analyzed, leaving an accurate, detailed map where the weed existed. From there, the WDNR outfitted a drone with a spraying mechanism and developed a program for the drone to disperse the treatment where the weed was located. In the end, the drone was more accurate than ground and helicopter sprayers, spraying only the weed and avoiding unaffected areas.

Drone mapping has also been used to monitor wildlife. Thermal sensors are attached to the drone so as it flies over an area, it picks up heat signatures that show where certain wildlife is present. Software is used to create a map of these locations, allowing scientists to focus their research on areas where the studied species was found.

While drones will definitely allow us to scan the Gobi Desert faster than trekking on foot, it would still help if we had an idea where in those 500,000 square miles we might find our cryptid.

THE SAXAUL PLANT

Locals have reported sightings in the southern portion of the desert in the hot and rainy months of June and July. In addition, encounters often occur in the vicinity of specific vegetation—the saxaul plant.

The saxaul plant is one of the few plants capable of surviving in the Gobi. It is a small (6- to 12-foot-tall), crooked, gnarly tree with branches that bend in odd shapes due to the high winds. It's extremely drought-resistant, as its bark absorbs water and then stores it. Mature saxaul trees have an extensive root system across and below the ground, where the roots search for moisture.

With a better understanding of *where* (south near saxaul plants!), *when* (June and July!), and *how* (drones!) to look for this cryptid, we can now formulate our search protocol.

MONGOLIAN DEATH WORM
SEARCH PROTOCOL

The Mongolian Death Worm search protocol is based on the hypothesis that this cryptid is an animal that predominantly lives underground near saxaul trees in the Gobi Desert.

1. Obtain a drone with cameras, sensors, and software capable of detecting the saxaul plant.

2. Program flight path for the southern area of Gobi Desert in either June or July (time when worm has been spotted).

3. Fly the drone over area to map presence of saxaul plant (tree present near worm sightings).

4. Analyze the locations of saxaul plant.

5. Obtain a drone with cameras, sensors, and software capable of detecting wildlife (worms, snakes, legless lizards).

6. Program flight path over areas where saxaul plant is present.

7. Analyze data for creatures resembling Mongolian Death Worm.

8. Follow up with field research at relevant locations and locate Mongolian Death Worm!

OTHER WORM- AND SNAKELIKE CRYPTIDS

MINHOCÃO

The Minhocão is reportedly a large burrowing creature/giant earth-worm/snake up to 150 feet long (about as long as 10 cars) and 15 feet wide. It's covered in thick, bony, armored skin or scales and produces trenches or furrows 3 to 10 feet wide. Its skin is as thick as pine tree bark and covered with scales like an armadillo.

Witness accounts date back to the mid-1800s in Brazil, where one of the creatures plowed up land surrounding a farm and another dug trenches deep enough to divert a tributary of a river. Later, one was reported to have died after becoming trapped in a rocky cleft near Uruguay. Other witnesses indicate they saw it plow furrows before burrowing underground.

TATZELWURM

Reports of this cryptid vary in description, but the most common is of a snakelike creature between 2 and 6 feet long. It has two front legs, is completely covered with scales, and has a face that resembles a cat. It's believed to be dangerously venomous and able to kill a human instantly with its bite. It's also been reported that it breathes poisonous fumes.

Sightings have occurred in south central Europe (Switzerland, Bavaria, Austria) for more than 100 years, continuing in the 2000s.

TSUCHINOKO

The tsuchinoko is a snakelike cryptid living in the mountainous regions of western Japan. It looks like a very wide common snake and is between 12 and 31 inches long with large, platelike scales running down its body. It reportedly has fangs and venom similar to common snakes. This cryptid has been sighted in fields, forests, and swamps. In 1969, a farmer claimed he caught one, but instead of saving it for research . . . he ate it. Yum?

SHUNKA WARAK'IN

The Ioway, a Native American Siouan people, are credited for giving the Shunka Warak'in (SHOON-kah wah-rahk-KEEn) its name, which translates to "carrying-off dogs." Other names used to reference this hyena-like, wolflike animal with black hair, a sloping back, and high shoulders include Ringdocus, the Rocky Mountain hyena, the Beast, and Guyasticutus.

With sightings in Canada, Montana, Nebraska, Iowa, and Illinois, the Shunka Warak'in appears to prefer the colder plains areas of the United States versus the woodsy areas other cryptids call home. It's also distinguished itself as a cryptid that's been caught.

Wait.

If it's already been nabbed, then what are we doing here? As is true with most things in cryptozoology, it's not that simple.

THE LEGEND

The late 1800s was a time of exploration in America. Using horse-drawn covered wagons, people left the hectic, crowded East and traveled west to start anew. Indeed, it wasn't until late 1889 that North Dakota, South Dakota, and Montana became part of the United States.

The Hutchins family was one of many who made this trek. Eventually settling in the lower part of Montana, Ross Hutchins established the Hutchins Ranch, where, one night, he encountered a strange animal chasing his grandmother's geese. He described it as a nearly black wolflike beast with high shoulders and a sloping back. When the same animal visited again later that same year, Ross's grandfather shot it and then donated the creature to a man who oversaw a grocery store and museum in Idaho.

The store owner mounted and displayed the unknown animal, calling it a Ringdocus. More than 100 years later, in 1995, Lance Foster, a member of the Iowan Nation, advised that descriptions of the Ringdocus resembled the Shunka Warak'in, a beast from the same parts that had been known to sneak into villages and carry off dogs.

If you feel like looking at this cryptid up close, the Ringdocus that Hutchins found is currently on display at the Madison Valley History Museum in Ennis, Montana. The animal's coat is dark brown, almost black, with lighter tan areas and a faint impression of stripes on its side. Its overall shape resembles a wolf, but it has an un-wolflike and more hyena-like sloping back. It is about 28 inches high at the shoulder and measures 4 feet long, excluding the tail.

MODERN SIGHTINGS

More recent potential sightings of the Shunka Warak'in place it in Alberta, Canada (which shares a border with northern Montana), where, in 1991, a hyena-like animal was spotted.

In December 2005, also in Montana, witnesses reported a wolf that was shot after killing more than 120 livestock in a period that spanned a year. Upon examination, the wolf seemed like it might not be a wolf at all. For one, the color was off. The creature had shades of orange, red, and yellow in its fur instead of the usual grays, browns, and blacks. Its feet were small, and it had a pointed face, which is uncharacteristic of wolves. Its perfect teeth were also unusual, as wolves often have broken teeth due to chewing bones. DNA testing results were initially inconclusive, revealing only some similarity to coyote DNA. Subsequent testing indicated a type of domesticated wolf or hybrid not found in the wild.

Those results are similar to the ones of a 2018 animal skulking through Montana. This creature, also caught, mystified witnesses as to whether it could be a wolf given the small size of its paws, its short teeth, big ears, and longer front claws. DNA ultimately concluded the animal was a gray wolf despite its unusual characteristics.

Finally, we have two reports of a cat-dog-hyena creature. First, in 2008, a witness in Illinois encountered a very large creature hunkered down like a cat—though it looked more like a dog and its front haunches resembled a hyena. Second, in 2019, a witness reported seeing a hyena lurking about in Colorado.

So, we have sightings of hyena-like and wolflike animals in Montana and the surrounding area for over 100 years. Are they all the same unknown species? Or are people slapping the Shunka Warak'in name on whatever strange-looking hyena/wolflike animal they're seeing?

After all, hyenas don't even live in North America. They're native to Africa and Asia only. That means it's unlikely the Shunka Warak'in is a *true* hyena, which further deepens the mystery of exactly what this animal is.

Wolf

Hyena

HYENAS AND WOLVES

If someone mistook a wolf for a dog, we'd likely think nothing of it. Both belong to the same family, Canidae, which also includes foxes and coyotes. Similarly, it can be easy to mix up some of the bigger cats in the Felidae (feline) family, which includes bobcats, cougars, lions, tigers, leopards, jaguars, panthers, cheetahs, lynxes, and ocelots.

You probably noticed hyena wasn't included under either category. That's because hyenas have a family all of their own—Hyaenidae— of which there are only four remaining species left in the world.

That being said, hyenas and wolves possess some similarities:

- ◆ **LENGTH**: 3 to 5 feet

- ◆ **HEIGHT**: 2 to 3 feet at shoulder

- ◆ **WEIGHT**: 80 to 150 pounds (though wolves can reach 190 pounds)

- ◆ **SPEED**: top speed in mid-30 mph range

- ◆ **DIET**: carnivorous

It can be hard to rely on color to distinguish one from the other, too. Even though hyenas are known to be sandy yellow or gray with spots and wolves are commonly gray, brown, and black, that's not true in all cases. The 2005 Montana wolf wasn't the common color of a wolf at all. When you add that some of these encounters are occurring at night or not in the best-lit locations and startled witnesses are catching just a glimpse, it's certainly not out of the question that witnesses could be using hyena and wolf to describe the same animal.

With so many similarities, any differences between the two become that much more significant. For example, the "sloping back" detail that some witnesses reported is a feature of hyenas, as a hyena's front legs are longer than its back legs. That height difference, along with the hyena's muscular shoulders, allows for a sloped-back appearance.

Another difference concerns their paw prints. Sometimes it can be difficult to distinguish a dog's print from a wolf's print because they're both canines. Since a hyena isn't part of the same family, though, their tracks are different. That means if we're able to examine the Shunka Warak'in's paw prints, we may gain information on what kind of animal it is.

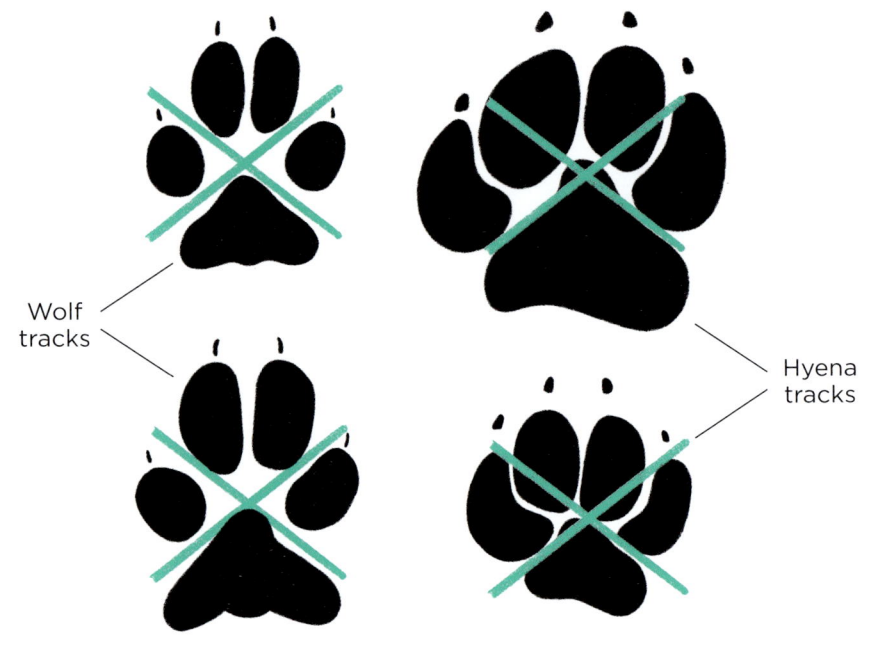

Wolf
tracks

Hyena
tracks

TRACKING AND FOOTPRINT IDENTIFICATION

Tracking animals via their paw prints is hardly a new phenomenon. Scientists, hunters, and wildlife enthusiasts have studied animal tracks for years, and the fundamentals remain the same. Identifying what animal left a certain paw print in the ground (or snow, mud, dirt, etc.) involves analyzing:

◆ the size of the track

◆ the number of toes

◆ whether the track has nails/claws

◆ the size of the front and back feet

Hyenas and wolves have four toes, and they also have nonretractable claws, meaning their paw prints will include a claw mark. Other features, though, are different.

WOLF TRACKS

Wolves leave very large prints, as much as 4 inches wide and 5 inches long, with the front track being a little bit longer and wider than the back print. Each paw has a heel pad and four toes, each with a claw, that are close together and point forward. A wolf's toes will all be the same size. One trick to identifying a wolf's paw print (or any canine, such as dog, fox, or coyote) is that the spacing between the toes and heel pad allows you to draw an X through the print.

HYENA TRACKS

The footprints of a hyena usually measure about 3.5 to 4 inches long for the front paw, with the back feet a tad smaller at 3 to 3.7 inches long. Each paw has a heel pad and four toes that fit very closely together. The back of the foot pad is also angled. Based on these characteristics, an X cannot be easily drawn.

FOOTPRINT IDENTIFICATION TECHNIQUE (FIT)

Scrounging around for tracks, which may be only partially present or otherwise obscured, isn't easy and can take loads of time. Thanks to the footprint identification technique (FIT), however, identifying animal tracks has become easier.

FIT was developed by a company called WildTrack, which was founded by two scientists committed to protecting endangered species. Frustrated with locating hard-to-find animal tracks of endangered rhinos, the scientists turned to technology to do the work for them. They took measurements and photos of rhino footprints and then uploaded them into software that took an additional 100+ measurements of a single print. Their database allowed AI to learn which prints belonged to which animal.

In the end, FIT can be more than 90 percent accurate in not only identifying which animal left a footprint but also in identifying that specific animal's sex and, sometimes, age-class. Scientists can then use that information to learn how many of a certain species visit a location and track individual animals. So far, some of the animals studied and identified using FIT include rhinos, tigers, cheetahs, pumas, and polar bears.

How can FIT help us? Once a database of hyena and wolf prints exists, all we need to do is snap a photo of the possible Shunka Warak'in's paw print, and FIT will tell us what species it is. If we find multiple sets of tracks that appear similar, FIT can even tell us if different individual animals left those tracks.

And, most importantly, if our collected print doesn't match *any* animal, then we may indeed have found a new species, which means it's time to check its DNA.

AIRBORNE DNA

In the chapter discussing the Beast of Busco, we examine environmental DNA (or eDNA) and how analyzing DNA in water samples can tell us what species visited that specific water source, in turn helping us pin down an animal's home. Unfortunately, eDNA is not as effective for monitoring the presence of terrestrial animals (animals living on land). If a land animal never dips its paw in or drinks from the stream being sampled, it's unlikely its DNA will be detected.

This is why recent studies sampling airborne environmental DNA are so crucial. The process of analyzing airborne DNA is very similar to that of eDNA. Instead of sampling water from a lake, stream, or water body, air is collected from filters placed at precise locations. Scientists then use an assortment of instruments to extract DNA from the filtered air. So far, airborne DNA has led scientists to successfully identify mammals, reptiles, and birds in the area where the air was collected.

As with eDNA, there are limits to what airborne DNA can tell us. Animals in closer proximity to the air collector or present in higher numbers are more likely to be detected. But placing a sampler in an area where our potential Shunka Warak'in (or its tracks) was recently spotted may lead to collecting its DNA so that we can conclusively identify its species or give scientists the ability to confirm it as a new species.

SHUNKA WARAK'IN
SEARCH PROTOCOL

Our Shunka Warak'in search protocol is based on the hypothesis that this cryptid is an unknown four-legged animal that resembles a wolf and/or hyena, so using wildlife science associated with studying land animals will help us pinpoint locations where it may be hiding.

1. Wait for new witness encounters of Shunka Warak'in.

2. Examine sites for animal tracks.

3. Take measurements and photos of tracks.

4. Implement FIT to analyze tracks.

5. Place airborne DNA samplers in locations where tracks are found.

6. Extract DNA from air samplers to determine species in the area.

7. Use FIT and DNA to identify Shunka Warak'in!

OTHER WOLFLIKE CRYPTIDS

BEAST OF BRAY ROAD

This creature has been described as a large animal resembling a wolf. It's been seen moving on all fours or solely on its hind legs, leading to werewolf comparisons. Estimates put it at 7 feet tall on its hind legs and 2 to 4 feet tall on all fours. Sightings initially occurred around Elkhorn, Wisconsin, which is about 40 miles southwest of Milwaukee, in 1936.

WAHEELA

This cryptid is known to roam the Arctic regions of Alaska and Canada. Witnesses report an enormous white wolf, much larger than normal wolves. Some have described an animal about 3½ feet tall at the shoulder with shaggy fur and short legs. Another reported a silvery white animal 5 to 6 feet tall on all fours that weighed 400 to 500 pounds and had a tail the size of almost half a car.

The waheela is said to bear resemblance to the Shunka Warak'in, though its white shaggy fur is a distinctive feature.

SKUNK APE

Phew! Something stinks in here. I know it's not me. If it's not you, maybe we need to look around. Perhaps a Skunk Ape lurks about!

As you may have deduced, this cryptid has been described as an apelike creature that emits a foul odor.

THE LEGEND

The Everglades lies at the southern tip of Florida. Spanning thousands of square miles, it's a natural region of tropical wetlands consisting of swamps, marshes, and tall grasses.

This environment is home to more than:

- 40 species of mammals

- 360 species of birds

- 300 species of fish

- 50 reptiles

- and . . . 1 ape?

Reportedly between 5 and 7 feet tall with red, dark brown, or black hair, this ape emits an unpleasant smell that has been compared to that of a skunk, rotten eggs, decaying corpses, or poop. While it bears some resemblance to the creature called Bigfoot, witnesses describe the Skunk Ape as an actual ape as opposed to a tall, large, hairy man.

Generally, apes rarely walk on two legs, though they are perfectly capable of doing so. In fact, a male gorilla named Louis at the Philadelphia Zoo was known to routinely strut across his enclosure on two legs, especially when he was holding food in his arms. According to his zookeepers, Louis did not like walking on all fours when the ground was muddy because he didn't like getting his hands dirty!

STINKY ENCOUNTERS

Early encounters of the Skunk Ape include a 4-year-old girl who spied a hairy creature in her backyard in 1947 and a large ape invading a campsite in the Big Cypress Swamp in 1957. In the 1970s, the Skunk Ape was spotted in the Big Cypress Swamp again. This time, witnesses made plaster casts of the footprints, ultimately concluding that, based on the different sizes, *three* Skunk Apes had visited their campsite.

Skunk Ape sightings continued through the 1990s and 2000s, all consistently describing a 5- to 7-foot hairy ape. One woman added that the orangutan-like creature stank and made *woomp* noises in her backyard. In some instances, witnesses examined footprints, many of which revealed an opposable big toe, which is common in apes.

THUMBS UP!

If you hold out your hand in front of you, you are able to move your thumb so that it can touch each of the other fingers on that hand. This maneuverability ("opposable thumb") allows you to grasp things, such as tools, utensils, etc. Good thing, too. Without an opposable thumb, how would you hold your smartphone?

Some animals, like opossums, monkeys, and apes, also have opposable thumbs. Many animals, though, have something we humans do not: opposable toes. Extend your foot and try to use your big toe to touch your other toes.

You can't.

Opposable toes allow animals to grasp objects with their feet as they (and we) are able to do with opposable thumbs on our hands. It's how monkeys can swing from branch to branch and climb and eat at the same time.

On other occasions, witnesses attempted to photograph or film the Skunk Ape. Some were more successful than others. From a distance, grainy footage of a hairy being is insufficient evidence upon which to conclusively identify an animal, much less a new species like the Skunk Ape. But one photo depicted a monkey-like animal that possessed features similar to an orangutan, and video from a fisherman taken at Lettuce Lake Park in Tampa in 2015 suggested some kind of primate lurking in the trees.

What makes sightings of a Skunk Ape so interesting in Florida is that apes and monkeys are not native to North America. Instead, apes (chimpanzees, bonobos, orangutans, and gorillas) are found in the tropical forests of Africa and Southeast Asia, while monkeys (mainly primates with a tail) call Asia, Africa, and Central America home. That being said, it's not like apes and monkeys *can't* live in North America. Florida's humid climate is especially fitting, but that would mean someone would first have to bring the animal here.

MONKEYING AROUND

Let's go back nearly 100 years to the 1930s. A man by the name of Colonel Tooey is trying to get more visitors to take his boat tours, which run along the picturesque Silver River in Ocala, Florida. Monkeys, he decides, will do the trick. He buys six macaques—a species of monkey that lives in North Africa and Asia—and sets them on a small island in the middle of the river.

Macaques are about 1 to 2 feet tall with arms and legs of the same length. They usually have brown or black fur and are considered highly intelligent. In addition, did you know they're great swimmers?

Don't feel bad—neither did Mr. Tooey. Supposedly, as soon as he placed those six macaques on the small island, they swam for their escape, disappearing into the surrounding forest. Undeterred, Mr. Tooey tried again and brought in six more macaques, which, like their predecessors, preferred to live elsewhere.

How much harm can 12 macaques do? Well, by the 1970s, over 80 macaques roamed Florida, and by 2015, 175 macaques were in Silver Springs State Park. Colonel Tooey wasn't the only person bringing monkeys to Florida, either. At one point, wildlife officials authorized the removal of more than *1,000* monkeys

The macaques have also expanded their territory to other areas of Florida. Between 2009 and 2012, one "rogue" macaque was spotted at several locations throughout Tampa, which is more than 100 miles south.

MACAQUE MAYHEM

Macaques can be quite the troublemakers. In 2022, in Lopburi, Thailand, local macaques ransacked homes, shops, and vehicles in search of food. It became routine to see macaques grabbing phones, water bottles, keys, and cameras and pilfering from vendor carts along the streets. One group of macaques even took over an abandoned cinema and defended it from other bands of macaques and humans, preventing them from entering.

The macaque escapade proves that monkeys can survive in Florida, but there's no way a macaque can be confused with a 5- to 7-foot-tall Skunk Ape. So, the question is, what *other* primates have people brought to Florida?

Turns out that Florida is home to several primate breeding, research, and/or sanctuary facilities. Combined, these facilities have the ability to house more than 5,000 animals, including macaques, baboons, chimps, and orangutans. Given the intelligence of these primates, it'd be foolish to rule out an escapee—especially since primates escape from zoos and other facilities quite . . . regularly?

In 2022, a chimpanzee escaped from a zoo in Ukraine. In 2021, 24 macaques (of course it was macaques!) escaped their enclosure at a zoo in Germany. In February 2020, 3 baboons escaped from a hospital for medical research in Sydney, Australia, and 70 monkeys escaped from a zoo in Japan.

Also, let's not forget the infamous orangutan named Ken Allen, who escaped from the San Diego Zoo three times in 1985 and could be seen patrolling the zoo with other visitors and, on some occasions, fellow escapee orangutans who followed him. There's also an orangutan who picked the lock on his enclosure with a wire to escape an Omaha Zoo in 1968.

To sum up: the possibility the Skunk Ape is an escapee cannot be easily dismissed!

STINK, STANK, STUNK!

As for the stench that surrounds Skunk Apes, that's not out of the realm of possibility either. Silverback gorillas, for example, are known for their pungent odor. In 2014, scientists revealed results of a year-long study regarding gorillas and this odor. Specifically, the scientists researched whether gorillas were *intentionally* emitting foul odors in certain situations as a form of communication.

After all, many animals use sense of smell to communicate. Dogs, for example, sniff to collect chemical information from other dogs, and when dogs pee on a tree or fire hydrant, the purpose is to leave a sign of their identity and mark their territory to other dogs that pass the area.

Were gorillas doing something similar?

The study showed that the lead gorilla was, in fact, varying its odor depending on the situation. The gorilla sent stronger odor signals when in the presence of other gorilla groups or rival males and emitted milder odor signals in other situations, such as when several group members were near. Overall, as the intensity of the lead gorilla's interactions increased, the emission of extreme odors also significantly increased.

These studies with gorillas show that it's possible the Skunk Ape may also have the ability to deliberately emit its foul odor, which is just the kind of break we need to track it.

SCENT-DETECTION DOGS

A dog's sense of smell is tremendous. Containing 200 million to 250 million scent receptors (compared to a human's paltry 5 million), a dog's sense of smell is even able to outperform sophisticated lab instruments. In fact, in 2021, scientists published a study showing that not only can dogs detect specific scents but also changes in emotions when humans have various illnesses.

With such a specialized ability, it's no wonder that dogs are used to sniff out potential dangers, such as at the airport. But now, conservation scientists are deploying dogs to detect endangered and invasive species. Dogs have been used to track animals such as bears, owls, wolves, garter snakes, turtles, bees, and koalas as well as to point scientists to harmful plants.

In 2019 and 2020, Australia experienced the most catastrophic bushfires in the country's history. More than 42 million acres were destroyed, over half of which consisted of forest and bushland, and scientists estimate 3 billion animals were killed, harmed, or displaced. With koala populations being hit especially hard, researchers deployed scent-detection dogs to locate and rescue injured koalas among the scorched earth. After specialized training, including having the dogs track the scent of koala fur instead of just koala poop, dogs rescued over 100 koalas.

In the United States, scientists at Yellowstone National Park have utilized scent-detection dogs to find zebra mussels. Zebra mussels are tiny, freshwater mussels. Though they can grow to be 2 inches, they're commonly the size of a fingernail. But they reproduce in such large numbers that they can clog pipes and waterways. They also eat algae that aquatic animals need to survive. If allowed to spread, they can destroy habitats.

Of course, being so tiny, zebra mussels are hard to detect. This mollusk's larvae are about the width of two human hairs laid side by side, allowing them to squirm their way into tiny crevices of boats and travel to new waterways or water bodies. There, they start a new infestation.

Helping scientists, scent-detection dogs have successfully sniffed out zebra mussels on boats using the park's waterways—thereby allowing scientists to remove them before they can travel and spread to a new location.

Before you run home and command your dog to start sniffing away, though, know that not just any dog can serve as a scent-detection dog. Scent-detection dogs must have the requisite training on the particular scent they are intended to find. Trained properly, however, scent-detection dogs have a high success rate in identifying the specified scent.

CAPTURING AND CREATING SCENTS AND ODORS

To train a dog to find the Skunk Ape, we first need to introduce the dog to the Skunk Ape's scent so that the dog knows what it's searching for. Since we don't have access to a Skunk Ape (that's why we need our scent-detection dogs!), we'll have to "create" the Skunk Ape's scent.

A scent is a specific combination of molecules we can sense through our nose. Different combinations of molecules will produce different scents. Perfumes, candles, and air fresheners are designed to produce captivating scents, like pine trees, ocean breezes, or fruity flavors.

Unfortunately, witnesses haven't reported that the Skunk Ape smells like fresh pizza dough or a bouquet of roses. Re-creating those aromas in a lab would be fun. Instead, to create the Skunk Ape's scent for our scent-detection dogs, we'll have to use more stinky molecules.

For example, the specific molecule that gives rotten eggs their awful smell is hydrogen sulfide (sulfur + hydrogen), and a skunk's spray consists of thiol, a compound that also contains sulfur.

Since witnesses reported both rotten eggs and skunk odors during their encounters with the Skunk Ape, utilizing molecules associated with those odors to create a scent will give our dogs a good sense of what they're sniffing for. Of course, if a witness were to trap surrounding air in a sealed container during an encounter with a Skunk Ape, we could analyze it and have an even better idea what molecules are at work.

FACIAL RECOGNITION TECHNOLOGY

Once we have the Skunk Ape in our sights, we'll need to figure out what we're looking at. What kind of primate is this? Have we actually found a new species? Using facial recognition technology, we may only need just a good photo of its face to find out.

With the advancement of camera traps (discussed during our visit with Mothman), scientists can collect photos of animals in the wild without being present so as to not risk scaring them away. Analyzing a large collection of photos takes a lot of time, though, which is where facial programming enters the picture. Using AI, computers can be trained to analyze photos for the animal being researched, meaning scientists no longer have to spend hours upon hours doing it themselves.

Developing the program takes time, as there must be enough images of the animal for the AI to "learn" what it's looking for. A study on brown bears required collecting almost 5,000 images of over 130 different bear faces. In the end, by focusing on facial features (eyes, nose, ears, forehead), the computer was able to "find a face" of a brown bear 97.7 percent of the time and was 80 percent accurate in recognizing a specific bear.

In developing a program to track cattle, 135,000 images of 1,000 cattle were used. The computer ultimately had a 94 percent accuracy rate in picking out cattle from photos.

The good news is scientists already have access to several databases used to identify animals—including primates—in the wild. This means that if we can photograph our Skunk Ape, we can download it to a database dedicated to identifying primate facial features and learn what species it is. If the AI cannot find a match, well, then maybe we have discovered a new species!

SKUNK APE
SEARCH PROTOCOL

Our Skunk Ape search protocol is based on the hypothesis that this cryptid is an as-of-yet-unidentified primate with a distinct odor, so focusing on its scent will help narrow down where to search.

1. Create and/or capture the scent of the Skunk Ape.

2. Train scent-detection dogs to locate the Skunk Ape's scent.

3. Release scent-detection dogs into Big Cypress Swamp and surrounding areas where the Skunk Ape has been witnessed.

4. Install camera traps in locations where scent-detection dogs indicate they've detected the Skunk Ape's scent.

5. Feed photos into facial recognition software.

6. Work with primate sanctuaries and conservation scientists to determine the Skunk Ape's species or if we've discovered a new species!

OTHER APELIKE CRYPTIDS

FOUKE MONSTER/BOGGY CREEK MONSTER/SWAMP STALKER

Known by a few names, this cryptid could be the Skunk Ape's relative. Or perhaps the Skunk Ape's been traveling to Arkansas since the 1970s? It's been described as a smelly apelike creature about 7 feet tall with long, reddish-brown hair and big red eyes. After a couple sightings, footprints found appeared to indicate the creature had three toes.

UYAN

The Uyan is an unknown ape in Malaysia. It's about 3 feet tall, is covered with dark hair, and has thin arms and legs. This cryptid is supposedly followed by a pack of dogs.

ORANG PENDEK

Sightings of this creature in the jungles of the island of Sumatra, Indonesia, have occurred for more than 100 years. Translated as "short person," the Orang Pendek is a primate 2½ to 5 feet tall covered in short fur. Its color can be black, gray, reddish-brown, or even honey.

Due to its harsh landscape, Sumatra remains largely unexplored, meaning an undiscovered species of great ape is not out of the question. In fact, in 2017, scientists in Indonesia discovered a new species of orangutan that differs genetically, morphologically, and behaviorally from the previously known two species of orangutans.

HIBAGON

This cryptid lives in the forests around Mount Hiba in Japan. About 5 feet tall, it reportedly has dark hair and a chocolate-brown face, and it emits a foul odor.

CONCLUSION:

THE EXPEDITION CONTINUES . . . WITH YOU!

J ust because you've reached the end of this book doesn't mean the expedition is over. You can research cryptids on your own, study animals with similar characteristics, and read about the latest technology so you can create your own search protocols.

Many of the methods (from field research to developing computer programs) we applied in our quests here are used on a daily basis by conservation scientists around the world. If you enjoyed reading about how scientists monitor and protect animals and the technologies they use in doing so, perhaps a career in conservation science is for you.

A **wildlife ecologist** studies the relationships between living things and their environment. Their goal is to understand how certain animals interact with each other and their surroundings and devise strategies to manage or protect them.

A **conservation biologist** works to stop the accelerated extinction process and facilitate the recovery of endangered life-forms. They monitor environmental conditions, population sizes, and other important environmental health indicators and, through research and observation, establish plans for maintaining habitats and animal populations at sustainable levels.

A **wildlife biologist** studies the biology, behavior, and habitats of a variety of animal populations in the wild. They may evaluate the impact of commercial development, study disease transmission, develop land- and water-use plans, and/or focus on saving endangered species.

A **wildlife biological technician** assists other wildlife scientists with their research. Most biological technicians do not have advanced degrees, which makes it a great option if you're interested in scientific research but prefer not to spend additional years in school.

Biological technicians perform experiments, record data, and prepare reports of their findings.

A **zoologist** studies the origins, genetics, diseases, life progression, and behaviors of animals. They work to preserve habitats, and some focus on protecting endangered species.

A **wildlife computer scientist** uses coding tools to train computers to perform various tasks involved with wildlife research. For example, if a lab is studying a certain breed of dolphin, a computer scientist may develop an algorithm for a computer to identify that dolphin from photos taken from underwater cameras.

This list is just a start. There are many careers to choose from! The great thing about science is that it truly has something for everyone.

THE EVOLUTION OF SCIENCE

The other thing to remember is that science doesn't stand still. While I tried to be as accurate as possible in this book from the information available at the time, there's certainly a chance that, by the time you're reading this, that information will have changed. New discoveries and technological advances can disprove theories we once relied upon.

We saw that with the study on loggerhead turtles discussed in the Altamaha-ha chapter underwater observations disproved the scientific theory that nesting female turtles were loners! Similarly, as noted in the Kraken chapter, it wasn't until scientists were able to view a giant squid in the ocean depths that they learned the giant squid stalks its prey. When new science debunks old science, it's important to take that new information, learn from it, develop new theories, and then test those theories so we can learn even more.

As an example, I'll tell you one last tale, this one about the legendary Loch Ness monster, which actually served as the inspiration for this book.

SCIENCE CAN CHANGE

A.K.A. THE LOCH NESS MONSTER CAN EXIST—
NO, IT CAN'T—
YES, IT CAN—
NO, IT CAN'T—
YES, IT CAN

Way back when the first modern sightings of the Loch Ness monster occurred in 1933, some theorized that the creature was a surviving plesiosaur. A plesiosaur is a broad, flat-bodied marine animal with a long neck and small head that could be up to 49 feet long. It lived in the Jurassic period, i.e., the age of dinosaurs.

The scientific community largely scoffed at this theory, saying plesiosaurs went extinct more than 66 million years ago at the end of the Cretaceous period.

Then, in 1938, the coelacanth (SEE-la-canth), a fish that can grow over 6 feet long and weigh over 200 pounds, was discovered off the coast of Madagascar. Why is this important? Well, until that discovery, scientists believed the coelacanth, which also lived in the age of dinosaurs, had gone extinct 66 million years ago at the end of the Cretaceous period.

Apparently not. Somehow, the coelacanth survived. (Not the same coelacanth, mind you. There's not a 66-million-year-old fish out there. The *species* managed to survive.)

Loch Ness monster believers seized on the discovery of the coelacanth as proof a plesiosaur—another aquatic animal—could have also survived this extinction event.

"Not so fast!" the scientific community countered. Scientists explained that plesiosaurs lived in salt water, not fresh water, and therefore, no, it could not be the creature living in Loch Ness.

Makes sense. We discussed in the Beast of Busco chapter how some animals live in fresh water and some live in salt water. So, sure, if science tells us that plesiosaurs only lived in salt water, then, no, the Loch Ness monster could not be a leftover plesiosaur.

That's how things stood for over 80 years until July 27, 2022, when a group of scientists published their discovery of plesiosaur fossils in the Sahara in Morocco. Now, fossils are found all the time—it's what paleontologists do. But *this* discovery was especially significant because 100 million years ago, when the plesiosaur was alive, the Sahara was a *river* system. In other words, these scientists had found evidence plesiosaurs lived in fresh water, too.

This finding surprised everyone, including the scientists who made the discovery, and raised entirely new questions:

◆ What were plesiosaurs doing in freshwater systems?

◆ Could the plesiosaur be like the beluga whale, which can tolerate fresh and salt water?

◆ Was there a particular time or reason the plesiosaur invaded freshwater systems instead of staying in the ocean?

Of course, to Loch Ness monster enthusiasts, the plesiosaur fossil discovery supports their belief that Nessie could be or could have been a plesiosaur.

Who's right? Who knows! Perhaps the science will change again.

The more exciting question is, what's next? What new technologies will bring about the next scientific discovery, and what will that discovery be? Could we finally invent a way to more easily monitor the depths of the ocean, and if so, what will we find?

The Kraken and I anxiously await the answer.

How about you?

FURTHER READING

Beccia, Carlyn. *Monstrous: The Lore, Gore, and Science Behind Your Favorite Monsters.* Minneapolis: Carolrhoda, 2019.

Coleman, Loren, and Jerome Clark. *Cryptozoology A to Z: The Encyclopedia of Loch Monsters, Sasquatch, Chupacabras, and Other Authentic Mysteries of Nature.* New York: Simon & Schuster, 1999.

Halls, Kelly Milner. *Cryptid Creatures: A Field Guide to 50 Fascinating Beasts.* Seattle: Little Bigfoot, 2019.

Ocker, J. W. *The United States of Cryptids: A Tour of American Myths and Monsters.* Philadelphia: Quirk Productions, 2022.

ABOUT THE AUTHOR AND ILLUSTRATOR

KIM LONG is a children's book author and former attorney. She studied environmental management in college and environmental law in law school. Her love of science, nature, and animals is reflected in her books. While she has not yet spotted a cryptid, she keeps her eyes peeled at all times, waiting for Mothman to make a guest appearance as she bikes along Illinois's trails.

NICOLE MILES is an illustrator from The Bahamas who currently lives in West Yorkshire (UK) with her pet snake and human boyfriend. She loves any opportunity to inform and entertain with her illustrations.